AGR 6928

HIGH-LEVEL POWER ANALYSIS
AND OPTIMIZATION

HIGH-LEVEL POWER ANALYSIS AND OPTIMIZATION

by

Anand Raghunathan
NEC USA

Niraj K. Jha
Princeton University

Sujit Dey
NEC USA

KLUWER ACADEMIC PUBLISHERS
Boston / Dordrecht / London

Distributors for North America:
Kluwer Academic Publishers
101 Philip Drive
Assinippi Park
Norwell, Massachusetts 02061 USA

Distributors for all other countries:
Kluwer Academic Publishers Group
Distribution Centre
Post Office Box 322
3300 AH Dordrecht, THE NETHERLANDS

Library of Congress Cataloging-in-Publication Data

A C.I.P. Catalogue record for this book is available
from the Library of Congress.

Copyright © 1998 by Kluwer Academic Publishers

All rights reserved. No part of this publication may be reproduced, stored in a retrieval system or transmitted in any form or by any means, mechanical, photocopying, recording, or otherwise, without the prior written permission of the publisher, Kluwer Academic Publishers, 101 Philip Drive, Assinippi Park, Norwell, Massachusetts 02061

Printed on acid-free paper.

Printed in the United States of America

CONTENTS

LIST OF FIGURES	ix
LIST OF TABLES	xiii
PREFACE	xv

1. INTRODUCTION			1
1.1	Low power design		2
	1.1.1	The emergence of portable systems	2
	1.1.2	Thermal considerations	3
	1.1.3	Reliability issues	4
	1.1.4	Environmental concerns	5
1.2	Design abstraction and levels of the design hierarchy		5
1.3	Benefits of high-level power analysis and optimization		12
1.4	Book overview		15
2. BACKGROUND			17
2.1	Sources of power consumption		18
	2.1.1	Capacitive switching power	18
	2.1.2	Short-circuit power	20
	2.1.3	Leakage power	20
	2.1.4	Static power	21
2.2	Methods for reducing power and energy consumption		22

2.3	High-level design techniques		25
	2.3.1 Module selection		29
	2.3.2 Clock selection		29
	2.3.3 Scheduling		29
	2.3.4 Resource sharing		31
2.4	High-level synthesis application domains		33

3. ARCHITECTURE-LEVEL POWER ESTIMATION 37
   3.1 Analytical power models 38
   3.2 Characterization based activity and power macromodels 43
      3.2.1 Activity-sensitive power macromodeling 45
      3.2.2 Accounting for glitching power consumption 50
      3.2.3 Bit-level and cycle-accurate power macromodels 59
      3.2.4 Improving macromodel efficiency with statistical sampling 63
      3.2.5 Improving estimation accuracy using adaptive macromodeling 65
   3.3 Power and switching activity estimation techniques for control logic 67
      3.3.1 Controller power consumption 68
      3.3.2 Estimating glitching activity in the control logic 70
   3.4 Conclusions 78

4. POWER MANAGEMENT 81
   4.1 Clock-based power management: Gated and multiple clocks 82
      4.1.1 Automatic synthesis of gated-clock circuits 84
      4.1.2 Clock gating techniques for data path registers 86
      4.1.3 Clock tree construction to facilitate clock gating 89
      4.1.4 Power management using multiple non-overlapping clocks 90
   4.2 Pre-computation 93
   4.3 Scheduling to enable power management 95
   4.4 Operand isolation 97
      4.4.1 Guarded evaluation 98

		4.4.2	Operand isolation in the context of high-level synthesis	100
	4.5		Power management through constrained register sharing	102
	4.6		Controller-based power management	107
	4.7		Conclusions	114

5. HIGH-LEVEL SYNTHESIS FOR LOW POWER — 115
   5.1 Behavioral transformations — 116
      5.1.1 Enabling supply voltage reduction using transformations — 117
      5.1.2 Minimizing switched capacitance — 119
   5.2 Module selection — 126
   5.3 Resource sharing — 129
      5.3.1 Exploiting signal correlations to reduce switched capacitance — 131
      5.3.2 Exploiting regularity to minimize interconnect power — 133
   5.4 Scheduling — 135
      5.4.1 Effect of scheduling on peak power consumption — 136
      5.4.2 Effect of clock period selection on power — 137
   5.5 Supply voltage vs. switched capacitance trade-offs — 139
   5.6 Optimizing memory power consumption during high-level synthesis — 142
   5.7 Reducing glitching power consumption during high-level design — 145
   5.8 Conclusions — 153

6. CONCLUSIONS AND FUTURE WORK — 155

REFERENCES — 159

INDEX — 173

LIST OF FIGURES

1.1	Factors driving the need for low power design	2
1.2	Synthesis flow and levels of abstraction	7
1.3	High-level design flow: Levels of abstraction	8
1.4	Projected growth in RTL synthesis tool seats	9
1.5	High-level synthesis benefits: Case studies	11
1.6	Benefits of high-level power analysis and optimization	12
1.7	Design flows without and with high-level power analysis	14
2.1	Illustration of capacitive switching power: (a) CMOS inverter, (b) equivalent circuit for charging the output load capacitor, and (c) equivalent circuit for discharging the output load capacitor	19
2.2	Data-flow intensive design example: VHDL description of a 6th order Elliptic Wave Filter	25
2.3	Control-flow intensive design example: VHDL description of a barcode pre-processor	26
2.4	Data flow graph of the EWF example	27
2.5	Control flow graph of the barcode pre-processor example	28
2.6	Functional RTL VHDL description of the EWF example	32
2.7	Structural RTL implementation of the barcode pre-processor example	33

3.1	Estimating clock line capacitance	39
3.2	High-level power estimation flow using power macromodeling: (a) macromodel construction, and (b) power estimation	44
3.3	Relationship between word-level temporal correlation and bit-level transition activity [1]	47
3.4	Transition template for a 2-input subtracter with misaligned breakpoints [1]	48
3.5	The GCD RTL circuit	51
3.6	Activity profiles for data path signals in the GCD circuit	52
3.7	Glitching activity models for an 8-bit subtracter	56
3.8	Circuit used to compute the coefficients D_{01}, D_{10}, and D_{11}	58
3.9	Variation of energy consumption with input/output switching activity for an 8-bit carry-lookahead adder [2]	60
3.10	Peripheral capacitance model	61
3.11	Overhead for computing bit-level statistics	63
3.12	Statistical sampling to improve the efficiency of high-level power estimation	64
3.13	Adaptive macromodeling to improve the accuracy of high-level power estimation	66
3.14	Activity profiles for control signals in the GCD circuit	67
3.15	(a) Implementation of control signal $contr[2]$, and (b) generation of glitches at gate $G1$	71
3.16	Scatter plot of switching activity at control signals: RTL estimate *vs.* gate-level estimate	78
4.1	Gating clock signals to save power	82
4.2	Gated-clock FSM architecture	84
4.3	Deriving clock gating conditions for data path registers	86
4.4	Clock gating at multiple levels in the clock tree	88
4.5	Effect of clock tree structure on clock gating possibilities	89
4.6	High-level synthesis of multiple clock designs	92

LIST OF FIGURES   xi

4.7	(a) Original circuit, and (b) circuit after applying pre-computation	93
4.8	Input subset disabling through pre-computation	94
4.9	Scheduling to enable power management	96
4.10	Operand isolation	98
4.11	Guarded evaluation	98
4.12	Operand isolation during high-level synthesis: Scheduled DFG	101
4.13	Operand isolation during high-level synthesis: RTL circuit	101
4.14	A scheduled CDFG to illustrate execution of spurious operations	103
4.15	Switching activity in the functional units of *Design 1*	104
4.16	Switching activity in the functional units of *Design 2*	105
4.17	Eliminating spurious operations using dynamic variable rebinding	106
4.18	RTL circuit implementing the send process of the X.25 protocol	109
4.19	Control re-specification example (a) ALU and its multiplexer tree, (b) original control expressions and activity graph, and (c) re-specified control expressions and activity graph	110
4.20	(a) Multiplexer tree feeding a register, (b) original control expressions and activity graph for signal M(18), and (c) re-specified control expressions and activity graph	111
4.21	(a) Comparator and its multiplexer trees, and (b) activity graph used for re-labeling	113
5.1	Using transformations to enable supply voltage reduction [3]	118
5.2	Minimizing switched capacitance by reducing the number of operations in the DFG [3]	120
5.3	Minimizing switched capacitance by strength reduction [3]	121
5.4	Using differential coefficients to minimize word-length of multiplication operations [4]	122
5.5	Average switching activity at the output of a constant multiplier *vs.* constant value [5]	124
5.6	Minimizing switching activity using transformations	125
5.7	Activity reduction in a linear chain [5]	126

xii   HIGH-LEVEL POWER ANALYSIS AND OPTIMIZATION

5.8	Minimizing power consumption through module selection	127
5.9	Using multiple supply voltages to minimize power	129
5.10	Effect of resource sharing on the switching activity in a shared resource [6]	130
5.11	Exploiting regularity to minimize interconnect power: (a) non-regular assignment, and (b) regular assignment [7]	134
5.12	Effect of scheduling on peak power consumption	137
5.13	Effect of varying the clock period on power consumption	138
5.14	Supply voltage *vs.* switched capacitance trade-off	140
5.15	Loop transformations for optimizing memory size and number of memory accesses [8]	143
5.16	Mapping arrays to memories in order to minimize transitions on the address bus [9]	144
5.17	Alternative architectures that implement the same function: Effect of glitching	146
5.18	Example circuit used to illustrate the effect of data signal correlations on control signal glitches	147
5.19	(a) Effect of data correlations on select signal glitches, and (b) use of the consensus term to reduce glitch propagation	148
5.20	Multiplexer restructuring to enhance data correlations: (a) initial multiplexer network, (b) abstract 3-to-1 multiplexer, and (c) restructured network	150
5.21	(a) Example circuit, (b) multiplexer bit-slice with selective delays inserted, and (c) implementation of a rising delay block	151

LIST OF TABLES

2.1	Characteristics of data-flow and control-flow intensive applications	35
3.1	Capacitance coefficients for a 2-input subtracter [1]	49
4.1	Two variable assignments for the scheduled DFG shown in Figure 4.14	104
5.1	Bit-level correlations between input and output values of operations	132

PREFACE

This book addresses issues that lie at the confluence of two ubiquitous trends in VLSI design – the move towards designing at higher levels of abstraction, and the increasing importance of power consumption as a design metric.

Power consumption is one of the most important metrics used in evaluating electronic systems today. This is due to a variety of requirements, such as prolonging battery life in portable devices, reducing chip packaging and cooling costs, and reliability and environmental considerations. Increasing clock frequencies and system complexities only serve to increase the demand for reducing power consumption. In order to address power consumption concerns, it is necessary to develop power estimation and reduction tools at each level of the design hierarchy.

The move towards designing at higher levels of abstraction is motivated by the growing complexities of electronic systems, shrinking product cycle times that require faster time-to-market, and the emergence of high-level design tools that support validation, analysis, and automatic synthesis starting from architectural and algorithmic (behavioral) design descriptions. Most research and development work in the areas of power analysis and optimization have addressed these problems at the lower (transistor and logic) levels of the design hierarchy. However, it is important to consider power consumption as a design metric at the higher levels of the design hierarchy for several reasons. Several studies have shown that large power savings are obtained through architectural and algorithmic trade-offs, which often far exceed the power savings obtained through lower-level optimizations. Tools that provide feedback about the power consumption of a design, given an architectural description, enable power budgeting decisions to be taken early on in the design flow, avoiding late surprises and possibly expensive design iterations.

The aim of this book is to describe techniques and tools that can be used to perform power analysis and optimization at the behavior and architecture levels. The book covers architecture-level power estimation techniques, power management during high-level design, and high-level synthesis techniques for minimizing power dissipation.

The authors would like to acknowledge their colleagues and students who helped shape their ideas and knowledge of low power design techniques; Rob Roy of Intel, Paul Landman of Texas Instruments, Allesandro Bogliolo of Stanford University, and Cheng-Ta Hsieh of the University of Southern California who provided additional material that helped with the description of their work; and Ganesh Lakshminarayana, Indradeep Ghosh, and Kamal Khouri of Princeton University, and Surendra Bommu of NEC C & C Research Laboratories, who read drafts of the chapters and suggested changes that improved the quality of the book.

ANAND RAGHUNATHAN

NIRAJ K. JHA

SUJIT DEY

This book is dedicated to our parents
Parthasarthy and Manjula Raghunathan
Chintamani and Raj Kishori Jha
Tarun and Kaberi Dey
and our spouses and children,
Shubha, Naina, Ravi, and Promit

HIGH-LEVEL POWER ANALYSIS AND OPTIMIZATION

1 INTRODUCTION

This chapter describes the factors driving the need for low power design, such as the growth in the portable electronics market, thermal considerations in very large scale integration (VLSI), circuit reliability issues, and environmental considerations. It introduces the commonly used levels of design abstraction, with an illustration of a typical high-level design flow, and an analysis of the benefits of incorporating high-level power estimation and optimization techniques into the design flow. The chapter concludes with an overview of the remaining chapters of the book.

2   HIGH-LEVEL POWER ANALYSIS AND OPTIMIZATION

1.1 LOW POWER DESIGN

The need for low power design is driven by several factors which are summarized in Figure 1.1. These factors are described next.

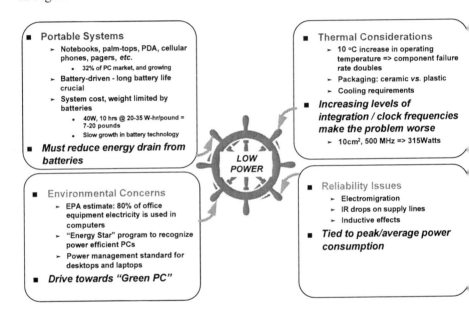

Figure 1.1. Factors driving the need for low power design

1.1.1 The emergence of portable systems

The emergence of portable or mobile computing and communication devices such as laptop and palmtop computers, cellular phones and pagers, wireless modems and network access cards, handheld video games, *etc.*, is probably the most important factor driving the need for low power design. Portable computers already account for a large portion of the personal computing market, and the demand for these devices is projected only to grow in the future. According to DataQuest Inc., the growth of the worldwide handheld electronics market in 1995 over 1994 was 24%

and the projected compound annualized growth rate for the mobile computing sector of the personal computer (PC) industry between 1995 to 2000 is 18.9%, compared to an estimated 16.9% for the entire PC market. Portable devices are battery-driven, and the life of the battery is a very critical parameter in the evaluation of such products – indeed, the commercial success of such a product depends significantly on its weight, cost, and battery life. Unfortunately, the advances in battery technology have not kept up with the growth in energy consumption requirements of the various system components, accentuating the need for low power design. Today's typical Nickel-Cadmium (Ni-Cd) batteries offer energy densities of around 23 Watt-hours/pound [10, 11], which means that providing 10 hours of operation to a device that consumes 20W of operating power would require a battery weight of 8.7 pounds. Newer battery technologies such as Nickel-Metal Hydride (Ni-MH) have capacities of 35-40 Watt-hours/pound. However, for the above scenario, the weight of the battery would remain significant. Thus, the cost and weight of the batteries become bottlenecks that prevent the reduction of system cost and weight unless efficient low power design techniques are adopted. For most portable devices, the power consumed in integrated circuit components is a significant and increasing portion of the total system power consumption [12]. Thus, the development of low power VLSI design methodologies and tools is of paramount importance.

1.1.2 Thermal considerations

The power consumed from the supply by integrated circuits is dissipated mostly in the form of heat. Efficient heat dissipation or cooling techniques are needed in order to maintain the chip's operating temperature within permissible levels. Failure to do so leads to circuit degradation and operating failures due to phenomena such as package-related failure, interconnect and junction fatigue, gate dielectric breakdown, and thermal runaway. It has been estimated that every $10°C$ increase in operating temperature causes the component failure rate to approximately double [13]. Low power design techniques lead to a reduction in cooling requirements,

4   HIGH-LEVEL POWER ANALYSIS AND OPTIMIZATION

which may lead to a reduction in packaging and cooling costs (*e.g.* it may be possible to use a plastic package instead of a ceramic package, or to eliminate the use of a cooling fan). Modern processors, like the Intel Pentium Pro (200MHz) which consumes 35 Watts, and the DEC Alpha 21164 which consumes 50 Watts, require expensive packages and cooling mechanisms. The power consumption in microprocessors has been growing, and is projected to grow roughly linearly in proportion to their die size and clock frequency [12]. On the one hand, high-performance systems will push packaging and cooling system limits, incurring steep cost overheads, while on the other hand, for high volume, low performance, and low power products, even a slight increase in per unit packaging and cooling costs will translate into large revenue reductions.

1.1.3 Reliability issues

Several reliability and signal integrity issues that affect the operation of integrated circuits are tied to their peak and/or average power consumption. For example, high levels of currents in metal interconnect lines lead to a phenomenon called *electromigration* where there is a mass transport of metal atoms, leading to electrical cuts or shorts between metal lines [14]. Electromigration is a major concern while designing the power supply network. Hot-carrier effects in MOS transistors (charge trapping in the oxide and/or interface trap generation at the Si/SiO_2 interface resulting in a shift in the threshold voltage and degradation in transconductance and electron mobility in the channel) are related to switching rates of transistors [15]. The *resistive (I-R) voltage drops* on supply lines also affect reliable circuit operation by resulting in degraded performance, reduced noise margins, and increased clock skews. Excessive supply current transients cause *ground bounce*, which refers to the ringing (voltage fluctuation) in the power supply network due to the inductances of the package pins and bonding wires. Specialized reliability and signal integrity analysis tools such as Railmill from EPIC Design Technology, Inc. [16] are required to help address these problems. Reducing the circuit's peak and average power consumption typically has the beneficial side-effects of improving the circuit's reliability.

1.1.4 Environmental concerns

Concerns about the direct and indirect environmental impact of computers is another motivation for low power design. According to an estimate by the U.S Environmental Protection Agency (EPA), 80% of the total office equipment electricity consumption is due to computing equipment, a large part of which is due to such equipment consuming current even when unused [17]. This led to the launching of efforts such as the EPA's *Energy Star* program [17], which outlines requirements for power-efficient PCs. This resulted in power management standards for desktops and laptops alike [18].

1.2 DESIGN ABSTRACTION AND LEVELS OF THE DESIGN HIERARCHY

Electronic designs can be represented at several levels of abstraction such as a geometric description of the layout, a logic description, or an architectural description. The hardware design process is often performed by gradually refining or detailing the abstract specification or model of the design to lower levels of abstraction. Circuit models are also classified in terms of the *views* of the design that they provide such as the *behavioral* or functional view, the *structural* view, and the *physical* view. A behavioral view specifies the functionality of the design and may contain little or no reference to its structure. A structural view represents the circuit as an interconnection of elements or building blocks. A physical view represents a circuit as a set of geometric entities that are placed on a chip or a board. Synthesis tools can automatically convert or refine a design from a higher level of abstraction to a lower level of abstraction, or can convert a description that presents a behavioral (structural) view to a structural (physical) view. High-level or behavioral synthesis converts a behavioral view of a design (henceforth referred to as a behavioral description for simplicity) into a structural view at the architecture level (henceforth referred to as an architectural or register-transfer level (RTL) description). Logic synthesis converts a behavioral view of a logic-level model

6   HIGH-LEVEL POWER ANALYSIS AND OPTIMIZATION

(*e.g.* a finite-state machine, or a set of Boolean equations) into a structural view at the logic level. Physical or layout synthesis converts a structural view of a design into a physical view.

Figure 1.2 shows the levels of abstraction and the synthesis steps involved in refining a design from each level to the next lower level. At the system level, the design may be modeled as a set of abstract communicating processes or tasks, with no knowledge of whether the tasks are implemented in hardware or compiled into software running on an "embedded" processor. System-level synthesis involves partitioning the tasks into hardware and software, choosing the processor(s) that will execute the software, determining the hardware/software communication mechanism, *etc*. For example, the figure shows the partitioned system-level block diagram of a barcode scanner device. Upon receiving a start signal from the microprocessor, the camera scans the barcode image, and generates a scan signal to indicate that it is starting to scan, and a digitized video signal that represents whether the current pixel is black or white. The pre-processor is an application-specific integrated circuit (ASIC), which processes the video signal sequence from the camera, computes the width of each white and bar, stores them in a memory, and asserts the eoc signal to indicate end of conversion. After that, the microprocessor reads the bar widths from the memory and performs the desired function, *e.g.* looking up the price of the scanned item.

The software components of the system go through the software implementation or compilation flow, which is beyond the scope of this book. The parts of the system that are to be implemented in hardware are represented by behavioral or algorithmic descriptions, where the detailed cycle-by-cycle behavior of the design and its structure may not be specified. High-level synthesis converts a behavioral description into a structural RTL implementation that is described as an interconnection of macroblocks and random logic. The RTL representation of the design can be expanded into a technology-independent logic-level netlist. Macroblocks that have pre-designed layouts (custom designs or module generators) are directly fed to the layout synthesis step. Logic synthesis optimizes a technology-

INTRODUCTION 7

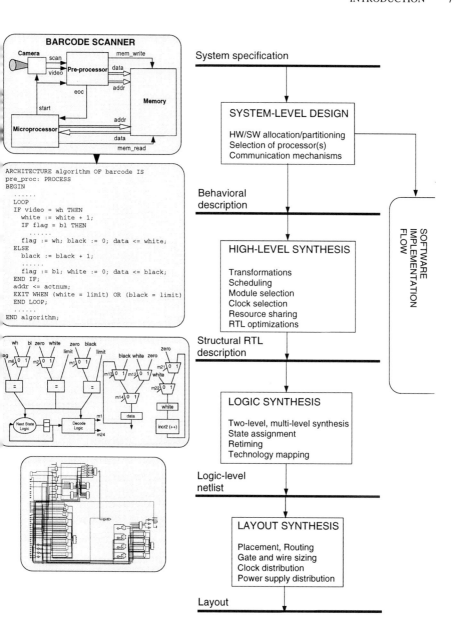

Figure 1.2. Synthesis flow and levels of abstraction

8 HIGH-LEVEL POWER ANALYSIS AND OPTIMIZATION

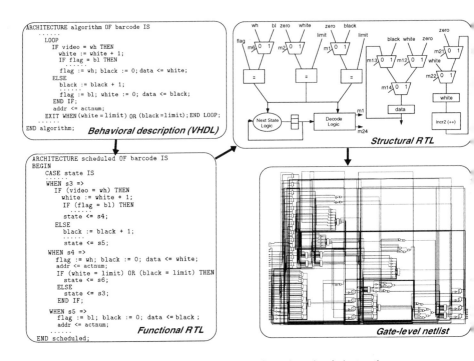

Figure 1.3. High-level design flow: Levels of abstraction

independent representation of a combinational or sequential logic function and maps it to a semi-custom technology library. The layout synthesis step performs tasks such as placement, routing, gate and wire sizing, and clock and power network generation, to result in a complete layout of the hardware, from which masks can be extracted for fabrication.

Figure 1.3 presents a closer look at the higher levels of design abstraction, that are the focus of this book. The top left portion of the figure shows a part of the behavioral description of the barcode pre-processor in the VHDL hardware description language [19]. Note that the behavioral description is very much like a high-level programming language in that it contains variables of simple data types like integers, operations on the variables, and dependencies among the operations.

INTRODUCTION 9

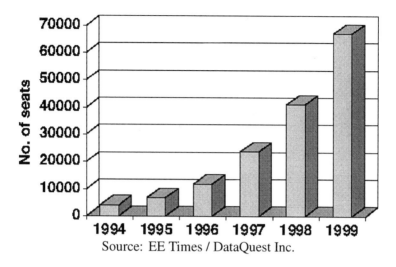

Source: EE Times / DataQuest Inc.

Figure 1.4. Projected growth in RTL synthesis tool seats

A behavioral description contains little or no information about timing, *i.e.* the clock boundaries or cycle-by-cycle behavior of the design is not fixed. The lower left portion of Figure 1.3 shows a part of the corresponding functional RTL description, where the cycle-by-cycle behavior of the design, *i.e.* the clock boundaries, are defined. A functional RTL description may be written manually or automatically generated from a behavioral description through the process of scheduling, which is one of the sub-tasks involved in high-level synthesis. The next step in the high-level design flow is to obtain the structural RTL description that implements the cycle-by-cycle functionality specified in the functional RTL description. The operations and variables of the functional RTL description are mapped to macroblocks such as arithmetic-logic units (ALUs), registers, and memory blocks, with other blocks like the control or random logic and multiplexers regulating the flow of data in the structure. The top right portion of Figure 1.3 shows part of the structural RTL implementation of the barcode pre-processor. A structural RTL design can be further refined into a gate-level netlist and then into a layout.

The trend towards designing at the higher levels

Recent years have seen a significant trend towards designs starting at the higher (architecture, behavior) levels of the design hierarchy. Figure 1.4 shows the projected growth in the total number of RTL synthesis tool seats from 1994 to 1999 (source: Dataquest, Inc.). The graph projects a growth from 11K in 1996 to 67K in 1999. The high-level synthesis segment of the electronic design automation (EDA) market is expected to display a compound annual growth rate of 75%, which makes it the fastest growing segment in the EDA market.

The factors driving the trend towards starting the design at higher levels of abstraction include:

- The growing complexities of integrated circuits make it difficult, time-consuming and error-prone to manually design at the lower levels. Automated synthesis tools allow the designer to capture and validate the design at higher levels, concentrating more on architectural trade-offs and less on the details of logic and physical design.

- The aggressive time-to-market requirements that drive most ASIC designs are pushing designers to adopt synthesis tools and design flows that enhance their productivity and cut design time.

- The use of high-level synthesis tools makes it possible to perform a thorough exploration of architectural trade-offs, which often results in large area, delay, and power savings.

Figure 1.5 reports the results of two industrial case studies that demonstrate the benefits of using high-level synthesis tools. These case studies indicate that the benefits that can be attained using high-level synthesis include reduced design turnaround times, and faster and smaller designs. Designing at the behavior level also leads to greatly reduced simulation and validation times, which further cut the design cycle.

	RTL Design		Behavioral Compiler (BC)		
Applications	Gates/Speed	Design Time	Gate/Speed	Design Time	BC Benefits
ATM Cell Scheduler	17,000 @ 40 MHz	6 Weeks	14,000 @ 40 MHz	2 Weeks	4 weeks faster to market
Graphics Processor Pixel Engine	35,000 @ 72 MHz	12 Months	30,100 80 MHz	3 Weeks	40% less latency
Satellite DSP	100,000 @ 20 MHz	20 Months	50,000 @ 20 MHz	9 Months	23X faster simulations
MPEG-2 Color Space Converter	9,800 @ 34 MHz	3 Months	10,000 @ 34 MHz	2 Weeks	10 weeks faster to market
Mass Storage Channel Controller	250,000 + 300k RAM @ 100 MHz	3 Weeks to modify FSM	250,000 + 300k RAM @ 100 MHz	2 Days	Automatically generate FSM

Source: Synopsys Inc.

	Manual RTL Approach	Behavioral Compiler Approach
Design Methodology	RTL VHDL, Top-Down	Behavioral, Top-Down
Lines of VHDL	23.8K	4.6K
Gate Count	90K	~50K
Throughput	28 cycles/frame	32 cycles/frame
Simulation Time	450 mins/frame	19 mins/frame
Man hrs Detailed Design	3360	1512
Design Duration	Four months	Three months

Source: Crevier et al., Raytheon Company, VIUF-95

Figure 1.5. High-level synthesis benefits: Case studies

12   HIGH-LEVEL POWER ANALYSIS AND OPTIMIZATION

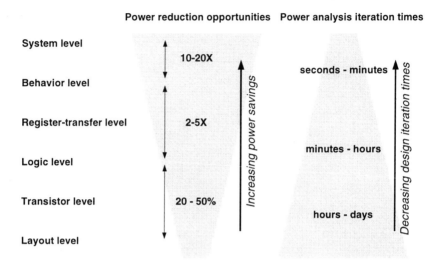

Figure 1.6.   Benefits of high-level power analysis and optimization

1.3 BENEFITS OF HIGH-LEVEL POWER ANALYSIS AND OPTIMIZATION

Given the increasing use of high-level design methodologies and the importance of low power design, this section examines the benefits of incorporating high-level power analysis and optimization tools into the design flow. Figure 1.6 lists the typical power reductions possible through the exploitation of power optimization opportunities at various levels of the design hierarchy. The figure also shows the typical design iteration times required to perform power analysis at the different levels of abstraction. Several studies have shown that the power optimization opportunities are significantly larger at the higher levels [12, 20]. System-level tradeoffs often yield an order of magnitude or more improvements in power. Algorithmic and architectural power management and optimization techniques can also yield large power savings. In comparison, the power savings obtained through logic and layout optimizations tend to be much smaller. With increasing levels of integration and operational speed, the power reduction requirements for most designs cannot be met by performing logic-level or transistor-level optimizations alone, pointing to the need to integrate power optimization techniques into the high-level design flow.

Power analysis tools are required in order to

- Validate that power budgets are met by the different parts of the design, and if not, identify the hot-spots in the design.

- Evaluate the effect of various optimizations and design modifications on power.

The use of high-level power analysis tools for the above purpose helps to greatly reduce the required design cycle. Figure 1.7 shows design flows without and with the use of high-level (architecture and system level) power analysis tools. In the absence of high-level power analysis tools, as indicated in the design flow of Figure 1.7(a), a power analysis iteration (*e.g.* to evaluate a design modification or alternative architecture) requires the designer to first synthesize and validate the functionality of a lower-level netlist, and then run a logic- or transistor-level power analysis tool to report power consumption. The combination of the large run times of lower-level power analysis tools, and the large time required to obtain and validate a gate- or transistor-level netlist make this methodology highly inefficient for exploring high-level design trade-offs, and infeasible for use in automatic high-level and system-level synthesis and optimization tools. In a design flow that uses high-level power analysis tools, such as the one shown in Figure 1.7(b), trade-offs at each level of the design hierarchy are supported by corresponding power analysis tools at the same level, leading to fewer and faster design iterations.

The reduced complexity of power analysis at the higher levels does not come without a penalty. The absolute accuracy of high-level power analysis tools tends to be lower than analysis tools at the lower levels of the design hierarchy. However, high-level power analysis tools are still very useful to guide high-level design

14 HIGH-LEVEL POWER ANALYSIS AND OPTIMIZATION

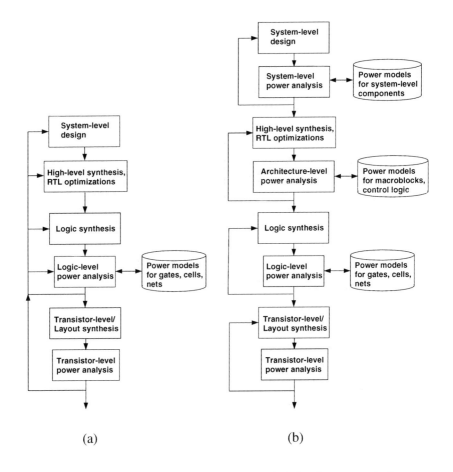

Figure 1.7. Design flows without and with high-level power analysis

trade-offs if their results provide relative accuracy (*i.e.* they are able to correctly predict whether a design modification will result in an increase or decrease in power consumption) and monotonicity (*i.e.* they are able to properly rank order a set of candidate designs in terms of power consumption) [21, 22]. With the use of high-level power analysis tools for exploring design trade-offs, the role of lower-level power analysis tools is limited to supporting lower-level optimizations, and verifying that the power budgets are met with a high level of confidence.

1.4 BOOK OVERVIEW

This book presents an overview of techniques to automatically perform power analysis and optimization for designs at the architecture or register-transfer (RT) level, and during high-level synthesis.

Chapter 2 provides the necessary background on power consumption in CMOS circuits, and high-level synthesis steps and techniques. The RT or architecture level is the design entry point for most designs today. The availability of a power estimation tool at this level of the design hierarchy eliminates need to synthesize gate- or transistor-level netlists and run low-level power analysis tools each time a high-level optimization or design modification is performed. Chapter 3 presents analysis techniques that can be used to estimate the power consumption of circuits described at the RT level. Power consumption in CMOS circuits is dominated by the dynamic component, which is incurred whenever signals in the circuit undergo logic transitions. In practice, a large fraction of the signal transitions that occur in a circuit are unnecessary. A popular class of power optimization techniques, called power management, are based on identifying and eliminating such unnecessary transitions. Chapter 4 describes several power management techniques that can be applied during the high-level design process. Moving higher up in the design hierarchy, it is often possible to achieve very large power savings by exploring the design space at the algorithm, or behavior level, and the relationship between the algorithm and implementation architecture. Chapter 5 describes high-level transformation techniques that can be applied to optimize designs at the behavior level, as well as techniques to perform high-level synthesis of a behavioral description to result in low power architectures. Chapter 6 summarizes the main conclusions that can be drawn from the material described in this book, and outlines directions for future work.

2 BACKGROUND

This chapter provides an overview of the sources of power consumption in digital CMOS circuits, general approaches to reducing power consumption, the sub-tasks involved in high-level design, and techniques commonly used to perform them.

2.1 SOURCES OF POWER CONSUMPTION

The sources of power consumption in digital CMOS circuits are summarized by the following equation:

$$P_{avg} = P_{sw.\,cap.} + P_{short-circuit} + P_{leakage} + P_{static} \qquad (2.1)$$

where $P_{sw.cap.}$ refers to the capacitive switching power, $P_{short-circuit}$ refers to short-circuit power, $P_{leakage}$ is the power consumption due to leakage currents, and P_{static} is the static power consumption.

2.1.1 Capacitive switching power

The capacitive switching power dissipation ($P_{sw.\,cap.}$) is caused by the charging and discharging of parasitic capacitances in the circuit. The computation of capacitive switching power is explained through the example of a CMOS inverter driving a load capacitor C_L that is shown in Figure 2.1(a). The output load capacitor C_L represents the cumulative effect of the parasitic capacitances associated with the nMOS and pMOS transistors (drain overlap and drain junction capacitances), the capacitance associated with the wiring internal and external to the inverter cell that is connected to the inverter's output, and the input capacitance presented by gates that this inverter drives.

Let us assume that the circuit is initially in a steady state, with the input at a logic value of 1, and the output at a logic value of 0. The output load capacitor C_L is discharged at this point. When the input undergoes a falling transition, the pMOS transistor turns on and the nMOS transistor turns off, as shown in the equivalent circuit of Figure 2.1(b). This leads to a charging current that results in C_L getting charged to V_{dd}. During this process, the energy that is drawn from the supply is $C_L.V_{dd}^2$, of which half is stored in the capacitor and the other half is dissipated in the pMOS transistor and interconnect. When the input undergoes a rising transition, the nMOS transistor turns on and the pMOS transistor turns off, as shown in Figure 2.1(c). This leads to a discharging current that flows through the

BACKGROUND 19

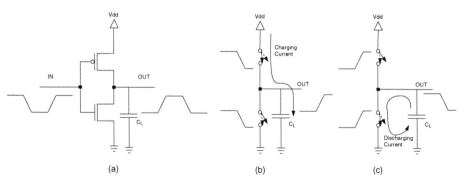

Figure 2.1. Illustration of capacitive switching power: (a) CMOS inverter, (b) equivalent circuit for charging the output load capacitor, and (c) equivalent circuit for discharging the output load capacitor

capacitor and the nMOS transistor, that results in the capacitor ultimately getting completely discharged. During this process, the $\frac{1}{2}C_L.V_{dd}^2$ energy that is stored in the output load capacitor gets dissipated in the nMOS transistor and interconnect. Thus, the capacitive switching power dissipated by the CMOS inverter shown in Figure 2.1(a) over a period of time $[0, T]$ can be computed using the following equation.

$$P_{sw.\ cap.} = C_L.V_{dd}^2.N_{0 \rightarrow 1}.\frac{1}{T} \qquad (2.2)$$

$N_{0 \rightarrow 1}$ is the number of rising transitions at the inverter's output, or equivalently the number of times C_L is charged, over the period of time $[0, T]$. Assuming that the inverter is part of a synchronous circuit running at a clock frequency f, and that the number of rising transitions ($N_{0 \rightarrow 1}$) is half the total number of transitions at the inverter's output, the above equation can be re-written as follows.

$$P_{sw.\ cap.} = \frac{1}{2}C_L.V_{dd}^2.N.f \qquad (2.3)$$

In the above equation, N is the average or expected number of transitions per clock cycle at the inverter's output, and is, henceforth, referred to as the *switching activity*. In most static CMOS technologies, the capacitive switching power accounts for a dominant part of the total power. As a result, most power estimation and optimization techniques targeting such technologies focus on reducing this component of a circuit's power consumption.

2.1.2 Short-circuit power

The short-circuit power ($P_{short-circuit}$) is caused by direct supply-to-ground paths that are created due to transients in signal values. Consider again the CMOS inverter shown in Figure 2.1. When the input changes from 1 to 0, there is a period of time when both the nMOS and pMOS transistors are conducting, leading to a short-circuit current being drawn from the supply. Assuming symmetric rise and fall delays and threshold voltages, the short-circuit power dissipation of a CMOS inverter can be approximated by the following equation [23].

$$P_{short-circuit} = K.(V_{dd} - 2V_T)^3.\tau.N.f \qquad (2.4)$$

where K is a constant that depends on the transistor sizes and the technology, V_T is the magnitude of the threshold voltage of the nMOS and pMOS transistors, τ is the input rise/fall time, N is the average number of transitions at the inverter's output, and f is the clock frequency. Short-circuit power dissipation can be controlled to a small portion of the total power by appropriate sizing of transistors and reducing the input rise/fall times to all the gates in the circuit. Short-circuit power dissipation is also reduced by scaling the supply voltage, and by reducing the switching activity at the gate outputs. Note that dynamic CMOS logic families, such as Domino CMOS [24], do not dissipate any short-circuit power.

2.1.3 Leakage power

The leakage power consumption ($P_{leakage}$) can be further decomposed into two components that are shown in the following equation [20]:

$$P_{leakage} = (I_{diode} + I_{subthreshold}).Vdd \qquad (2.5)$$

In the above equation, I_{diode} refers to the currents flowing through the reverse-biased diodes that are formed between the diffusion regions and the substrate. These currents are very small for current technologies, a typical value being 1 femto-ampere per device junction.

The term $I_{subthreshold}$ refers to the currents arising due to the fact that transistors that are "off" conduct some non-zero current. The expression for subthreshold current in the nMOS transistor of a CMOS inverter whose input voltage V_{in} varies between 0 and its threshold voltage V_{T_n} is as follows [23].

$$I_{subthreshold} = K.W_{eff}.e^{\frac{V_{in}-V_{T_n}}{S}} \qquad (2.6)$$

where K and S are constants that depend on the technology, and W_{eff} is the effective transistor channel width. The subthreshold power is small for current technologies. However, it increases for transistors with large channel widths, and more significantly at reduced threshold voltages.

Leakage power dissipation is especially important for devices that are in an idle state most of the time.

2.1.4 Static power

Static power consumption (P_{static}) is of importance for logic families such as pseudo-nMOS, where a gate consists of a single pull-up pMOS transistor and an nMOS network, in which there is a constantly conducting supply-to-ground path. In fully complementary CMOS circuits, static power consumption can result due to degenerated voltage levels at the inputs to a static gate, or due to selector- or bus-conflicts where multiple drivers attempt to drive a signal to different logic values. Such situations are undesirable, and are typically avoided through proper circuit design techniques.

2.2 METHODS FOR REDUCING POWER AND ENERGY CONSUMPTION

It is important to make a distinction between the terms power consumption and energy consumption which are often used interchangeably [25]. Battery-driven systems are limited by the amount of energy that can be supplied by the battery. Power is the rate at which energy is drawn from the batteries. Power dissipation is critical from cooling and packaging considerations, whereas energy consumed is important for battery life considerations. If a computation can be performed in such a way as to consume half the power it previously did (*i.e.* draw energy from the batteries at half the rate) but takes twice the time to complete, the energy consumed for the entire computation is no different from what it was before. Therefore, power savings and energy savings, in the strict sense of the terms, do not necessarily go hand-in-hand. However, given a fixed time for a computation to complete, power and energy consumption do vary proportionately.

The equations for the various components of power consumption presented in the previous section indicate that the parameters that can be varied to affect power as well as energy consumption are the supply voltage, the clock frequency, the switching activity per clock cycle at various signals in the circuit, and the parasitic capacitances. Optimizing power consumption invariably involves reducing one or more of these parameters. It is important to note that these parameters are not independent. It is necessary to take into account the interactions and trade-offs among these parameters to minimize power consumption.

Reducing the supply voltage (V_{dd}), which has a quadratic effect on the switching power consumption, is often the most fruitful way of reducing power. Supply voltage reduction, however, does not come without penalties. The delay of circuit elements increases according to the following equation [25].

$$delay = k . \frac{V_{dd}}{(V_{dd} - V_T)^2} \qquad (2.7)$$

where k is a constant. Thus, performance requirements impose a limit on the extent to which supply voltage scaling can be done. Reductions in the supply voltage also result in a reduction in the circuit's noise margins, making the circuit more susceptible to noise-related soft failures [20]. A practical consideration arises due to the requirement for standard supply voltages in order to enable easy integration of off-the-shelf components. The availability of high-efficiency DC-DC voltage converters for use at the chip and board levels eases this problem to a certain extent [26]. A common practice is to use separate supply voltages for the I/O circuitry and the chip's core.

Several approaches have been proposed to maximize the extent of voltage scaling possible. For example, scaling down the dimensions of devices along with the supply voltage compensates for the negative effects of supply voltage on performance. Another technique used to avoid the performance penalties of supply voltage scaling is to reduce the threshold voltage V_T of transistors [20]. However, decreasing the threshold voltage leads to a significant increase in the power consumption due to subthreshold and leakage currents, making it disadvantageous to lower the threshold voltage beyond a certain point. One solution that has been proposed to work around this problem is to dynamically vary the threshold voltages of devices, using lower threshold voltages when devices are active, and higher threshold voltages when they are not [27]. The use of faster logic and architectural blocks such as carry-lookahead adders instead of ripple-carry adders, Wallace multipliers instead of array multipliers, *etc.*, can be used to enable supply voltage reduction at the expense of switched capacitance [28, 29, 30]. The use of architectural parallelism through the replication of computational resources and the use of pipelining [31], performance-enhancing architectural transformations [3], and selection of efficient algorithms [32] have been used to enable supply voltage scaling without degrading performance. The use of multiple supply voltage implementations has been shown to reduce power as compared to implementations that use a single supply voltage [29, 33, 34, 35]. The idea is to use a lower supply voltage for circuit blocks that are not on the critical path, resulting in little or no

24 HIGH-LEVEL POWER ANALYSIS AND OPTIMIZATION

degradation of performance. Circuit design techniques that clamp or limit the voltage swings at signals internal to logic gates and memories to reduce power consumption are described in [36, 37].

The product of switching activity and physical capacitance, termed switched capacitance, is another parameter that can be minimized to reduce power consumption. Switched capacitance reduction techniques have been developed at all levels of the design hierarchy. At the technology level, the reduction of feature sizes and the use of low-parasitic process technologies result in a reduction of switched capacitance. Physical design techniques such as partitioning, placement and routing, transistor and wire sizing, transistor re-ordering, and clock tree construction can also be used to reduce switched capacitance [38]. Overviews of the various logic-level optimization techniques to reduce switched capacitance, including two-level and multi-level combinational circuit synthesis, technology mapping / cell selection, state encoding, retiming, automatic synthesis of gated clock circuits, and logic-level power management techniques are presented in [39, 40, 41]. At the architecture level, power management, efficient data representation and encoding, the use of multiple clocks, architectural transformations, memory segmentation, and high-level synthesis techniques have been used to reduce switched capacitance [12]. Algorithm-level optimizations, like power-driven algorithm selection, algorithm/architecture matching, global communications and memory access optimizations, and the exploitation of locality and regularity can lead to large reductions in switched capacitance. The use of software optimization techniques such as instruction selection, code generation, instruction scheduling, strength reduction, and architecture-specific compiler optimizations to reduce switched capacitance are described in [42, 43].

Reducing the clock frequency alone will result in power reductions, but it will not affect the amount of energy consumed for performing a specific computation and is, hence, not always a useful means of energy reduction. However, when the circuit is idle or needs to perform very little computation for significant periods of time, dynamically reducing the operating frequency helps eliminate unnecessary power as well as energy consumption. This technique is commonly used as a power saving strategy in microprocessors [18].

2.3 HIGH-LEVEL DESIGN TECHNIQUES

High-level (or behavioral) synthesis is the process of deriving a structural RTL implementation of a design that implements a given behavioral (functional or algorithmic) specification. Design metrics like area, performance, power, and testability could either be constraints or co-objectives during the synthesis process. Behavioral descriptions are usually written in a hardware description language

```
ENTITY digital_filter IS
  PORT (input:  IN  integer;
        output: OUT integer;
        go:     IN  bit;
        done:   OUT bit);
END digital_filter;

ARCHITECTURE behavior OF digital_filter IS
BEGIN
  PROCESS
    TYPE coef_arr is array (integer range 1 to 10) of integer;
    VARIABLE c: coef_arr := (0,1,2,3,4,5,6,7,8,9);  -- coefficient array
    VARIABLE s1, s2, s3, s4, s5, s6: integer := 0;  -- state vars
    VARIABLE x0, x1, x2, x3, x4, x5, x6: integer;   -- temp vars
    VARIABLE v1, v2: integer;  -- common subexpression vars

  BEGIN
    done <= '0';
    -- ready to read input, wait until input is ready
    WAIT until go = '1';

    x0 := input;
    x1 := s1;
    x2 := s2;
    x3 := s3;
    x4 := s4;
    x5 := s5;
    x6 := s6;
    v1 := x4 + x2 - (x0 - x1) * c(7);
    v2 := x5 * c(5) + x6 - x4;
    s1 := v1 * c(9) + v2 * c(9) + x1;
    s2 := x2 + x5 * c(4) - x5 * c(2);
    s3 := x4 * c(8) + x3;
    s4 := (x1 - x3 - x5) * c(6) + x4;
    s5 := v1 * c(7) + v2 * c(10) + x5;
    s6 := x1 * c(3) - (x5 * c(1) + x6);
    output <= s5;

    done <= '1';
    -- output is available,
    -- wait until output is read
    WAIT until go = '0';
  END PROCESS;
END behavior;
```

Figure 2.2. Data-flow intensive design example: VHDL description of a 6th order Elliptic Wave Filter

26 HIGH-LEVEL POWER ANALYSIS AND OPTIMIZATION

```
ARCHITECTURE algorithm OF barcode IS
BEGIN barcode: PROCESS
  BEGIN
    eoc <= false; memw <= false;
    data <= 0; addr <= 0;

    RESET_LOOP: LOOP
      LOOP EXIT WHEN start; END LOOP;
      LOOP
        LOOP EXIT WHEN scan; END LOOP;
        flag := wh; actnum := 0; white := 0; black := 0;
        eoc <= false;
        LOOP
          IF video = wh THEN
            white := white + 1;
            IF flag = bl THEN
              actnum := actnum + 1; memw    <= false;
            ELSE
              memw    <= true;
            END IF;
          ELSE
            black := 0; flag := wh; data <= white;
          ELSE
            black := black + 1;
            IF flag = wh THEN
              actnum := actnum + 1; memw    <= false;
            ELSE
              memw    <= true;
            END IF;
            flag := bl; white := 0; data <= black;
          END IF;
          addr <= actnum;
          EXIT WHEN (white = 255) OR (black = 255); END LOOP;
          EXIT WHEN (actnum = num) AND (white = 255); END LOOP;
          memw <= false; eoc <= true;
          LOOP EXIT WHEN start = false; END LOOP;
        END LOOP RESET_LOOP;
      END PROCESS;
    END algorithm;
```

Figure 2.3. Control-flow intensive design example: VHDL description of a barcode pre-processor

(HDL) such as VHDL, Verilog, or Hardware C. Tools that facilitate graphical design entry and facilitate automatic HDL code generation such as DesignBook from Escalade Corporation, are often useful in enhancing designers' productivity. Behavioral descriptions may contain simple abstract data types (ADTs) (such as integers, floating point numbers, enumerated types, *etc.*), arithmetic and logical operations on ADTs, control and iteration constructs, and sub-programs. Two example behavioral descriptions written in VHDL are shown in Figures 2.2 and 2.3. Figure 2.2 shows the specification of a 6th order Elliptic Wave Filter (EWF). This example belongs to a class of applications that are collectively referred to as the *data-flow intensive* or arithmetic intensive application domain, where most of the computations performed in the design are arithmetic operations such as addition, subtraction, and multiplication. Digital signal and image processing applications, and some multimedia applications fall under this category. In contrast, there is the *control-flow intensive* application domain, where designs contain significant control flow constructs like nested loops and conditionals. The barcode pre-processor example shown in Figure 2.3 could be classified as a control-flow intensive design. Note that the only arithmetic operation performed in the design is increment by

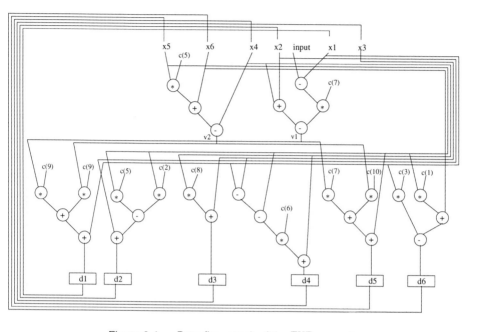

Figure 2.4. Data flow graph of the EWF example

1. In comparison, the design contains numerous nested loops and conditionals, and conditional data assignments. Data-flow and control-flow intensive designs have significantly differing power, area, timing, and testability characteristics, which are elaborated on later in this section. The behavior of the HDL description is validated through the use of simulation and/or formal verification techniques before proceeding with synthesis.

High-level synthesis tools typically compile a behavioral description into a suitable intermediate format. Depending on the application domain, commonly used intermediate representation formats are the *data flow graph* (DFG) and the *control flow graph* (CFG) [44]. A DFG represents operations in the behavioral description as vertices, and data dependencies between the operations as edges between the corresponding vertices. The DFG for the EWF example is shown in

28 HIGH-LEVEL POWER ANALYSIS AND OPTIMIZATION

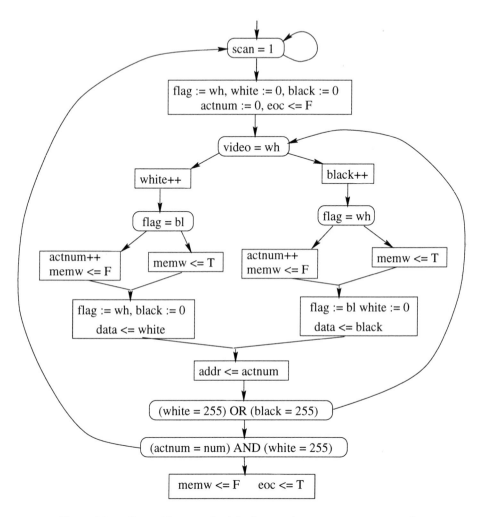

Figure 2.5.   Control flow graph of the barcode pre-processor example

Figure 2.4. The boxes labeled $d1, ..., d6$ are *delay elements*, *i.e.* they store results from one iteration of the computation for use in the next iteration. In a CFG, vertices represent operations, while edges represent control (or sequencing) dependencies between operations. The CFG of the `barcode` pre-processor example is shown in Figure 2.5. Composite representations such as the control-data flow graph (CDFG) [45] have been developed to capture the control and data flow into a single data structure.

The different sub-tasks in high-level synthesis are discussed next.

2.3.1 Module selection

Module selection refers to the process of selecting, for each operation in the CDFG, the type of functional unit that will perform it. The decision is trivial if there is only one functional unit template in the library that can perform each operation (*e.g.* the only functional unit that can perform addition is *ripple_carry_adder*). In order to fully explore the design space, however, it is necessary to have a diverse library of functional unit templates where multiple templates exist that are capable of performing each operation (*e.g. ripple_carry_adder, carry_lookahead_adder, carry_select_adder* for addition, *array_multiplier, wallace_tree_multiplier, pipelined_multiplier* for multiplication, *etc.*). Module selection can be viewed as analogous to the technology mapping problem in logic synthesis, where a network of logic elements is mapped to cells from a technology library. Researchers have proposed techniques to perform area and delay trade-offs using module selection [46, 47].

2.3.2 Clock selection

Clock selection refers to the process of choosing a suitable clock period for the controller/data path circuit. Several high-level synthesis systems do not explicitly perform clock selection during synthesis but allow the clock period to be determined after the complete implementation is obtained. The importance of judicious clock selection has been demonstrated in [48, 49, 50].

2.3.3 Scheduling

The process of scheduling determines the cycle-by-cycle behavior of the CDFG, *i.e.* it assigns each operation in the CDFG to one or more cycles or *control steps*. Each control step corresponds to a time interval equal to the clock period. Scheduling has been shown to affect the performance, area, power, and testability of the resulting implementation. Scheduling techniques are, therefore, classified based on the design metrics they accept as constraints and those that they attempt to optimize. For example, time-constrained scheduling and resource-constrained scheduling are two commonly used families of scheduling techniques. In time-constrained scheduling, the total execution time required by the implementation to process an input (often approximated as the number of control steps required) is a user-specified constraint that must be satisfied, while any combination of other design metrics such as power, area, or testability could be optimized. In resource-constrained scheduling, on the other hand, the resources available to implement the data path (*e.g.* the number of functional units of each type, registers, buses) are fixed, and the number of control steps is minimized.

The simplest scheduling algorithms are as-soon-as-possible ($ASAP$) and as-late-as-possible ($ALAP$) scheduling where operations are scheduled at the earliest (latest) possible control steps subject to satisfaction of their data dependency constraints. Both of these methods result in schedules that require the minimum number of cycles for acyclic DFGs, but these schedules typically require an excessive amount of resources. List scheduling is a constructive method that schedules operations one control step at a time onto a set of available functional units. A list of currently unscheduled operations is maintained. An operation from the list can be scheduled as soon as its data dependency constraints have been satisfied. Since multiple operations can have their data dependency constraints satisfied in the same control step, a priority function is used to choose among them. For example, the mobility (the difference between the $ASAP$ and $ALAP$ control steps of an operation) can be used as a priority function [51]. Sophisticated scheduling algorithms such as force-directed scheduling [52], simulated annealing based approaches [53], iterative-improvement based scheduling [54], integer-linear programming based approaches [55, 56, 57], path-based scheduling [58], loop-directed scheduling [59], wave scheduling [60], and several other algorithms have been proposed. Detailed descriptions of popular scheduling algorithms can be found in [44, 47].

The output of the scheduling process is a refined version of the behavioral description where the cycle-by-cycle behavior of the design is specified. Designs at this intermediate level are referred to as scheduled behavioral or functional RTL designs. The functional RTL VHDL description of the EWF example is shown in Figure 2.6. Note that there is a notion of a set of *controller states* and transitions among them. Each controller state corresponds to one or more clock cycles, and the operations in the behavioral description are assigned to the various controller states.

2.3.4 Resource sharing

Resource sharing or hardware sharing refers to the use of the same hardware resource (functional unit or register) to perform different operations or store more than one variable. The high-level synthesis tasks that perform resource sharing are *resource allocation* and *resource assignment*. The process of allocation determines the number of resources including functional units, registers and buses needed to implement the design. Assignment or binding refers to the process of mapping operations in the behavioral description to the allocated functional units, variables to registers, and data transfers to buses or multiplexers. When performing resource-constrained scheduling, resource allocation is performed before scheduling. Resource assignment or binding is typically performed after scheduling, since operations that need to be performed concurrently cannot share a functional unit, and variables whose values need to be stored simultaneously cannot share the same register. After functional unit and register allocation and assignment have been performed, the communication paths among the functional units and registers are generated as a combination of dedicated multiplexer-based (point-to-point) interconnects and shared buses. The functional units, registers, multiplexers, and buses together constitute the data path of the implementation. The scheduling and resource sharing information is used to construct a controller finite-state machine

32 HIGH-LEVEL POWER ANALYSIS AND OPTIMIZATION

```
PACKAGE types IS                                          t10 := x5 * c(2);
  TYPE states IS (s0, s1, s2, s3, s4, s5, s6, s7);        t12 := x4 * c(8);
END types;                                                t13 := x1 - x3;
USE work.types.all;                                       t19 := x1 * c(3);
                                                          t20 := x5 * c(1);
ENTITY digital_filter IS                                  WHEN s2 =>
  PORT (clk:   IN  bit;                                   t3 := t2 * c(7);
        input: IN integer;                                t5 := t4 + x6;
        output: OUT integer;                              t11 := t9 - t10;
        go:    IN bit;                                    s3 := t12 + x3;
        done:  OUT bit);                                  t14 := t13 - x5;
END digital_filter;                                       t21 := t20 + x6;
                                                          WHEN s3 =>
ARCHITECTURE scheduled OF digital_filter IS               v1 := t1 - t3;
BEGIN                                                     v2 := t5 - x4;
  PROCESS                                                 s2 := t11 + x2;
    TYPE coef_arr IS array (integer range 1 to 10) OF integer;   t15 := t14 * c(6);
    VARIABLE c: coef_arr := (0,1,2,3,4,5,6,7,8,9); -- coefficient array   s6 := t19 - t21;
    VARIABLE s1, s2, s3, s4, s5, s6: integer := 0; -- state vars         WHEN s4 =>
    VARIABLE x0, x1, x2, x3, x4, x5, x6: integer; -- temp vars           t6 := v1 * c(9);
    VARIABLE v1, v2: integer; -- common subexpression vars               t7 := v2 * c(9);
    -- temporary intermediate vars                                       t16 := v1 * c(7);
    VARIABLE t1, t2, t3, t4, t5, t6, t7, t8, t9, t10: integer;           t17 := v2 * c(10);
    VARIABLE t11, t12, t13, t14, t15, t16, t17, t18: integer;            s4 := t15 + x4;
    VARIABLE t19, t20, t21: integer                                      WHEN s5 =>
    VARIABLE state: states := s0; -- global controller state             t8 := t6 + t7;
                                                                         t18 := t16 + t17;
  BEGIN                                                                  WHEN s6 =>
    WAIT until (clk = '1' and clk'event);                                s1 := t8 + x1;
    CASE state IS                                                        s5 := t18 + x5;
    WHEN s0 =>                                                           output <= s5;
      done <= '0';                                                       -- indicate that output is available
      -- wait until input is available                                   done <= '1';
      IF(go = '1') THEN                                                  state <= s7;
        state <= s1;                                                     WHEN s7 =>
      ELSE                                                               -- wait until output is read
        state <= s0;                                                     IF (go = '0') THEN
      END IF;                                                            state <= s0;
    WHEN s1 =>                                                           ELSE
      x0 := input; x1 := s1; x2 := s2; x3 := s3;                         state <= s7;
      x4 := s4; x5 := s5; x6 := s6;                                      END IF;
      t1 := x4 + x2;                                                     END CASE;
      t2 := x0 - x1;
      t4 := x5 * c(5);                                                   END PROCESS;
      t9 := x5 * c(5);                                                   END scheduled;
```

Figure 2.6.   Functional RTL VHDL description of the EWF example

that configures the data path to perform the appropriate computation in each clock cycle. The data path and controller are together referred to as the structural RTL implementation. The structural RTL implementation of the barcode pre-processor example is shown in Figure 2.7. Techniques to optimize area, delay, and testability during resource sharing have been investigated [44, 47]. Chapter 5 presents resource sharing techniques that minimize power dissipation.

The structural RTL circuit that is produced as a result of high-level synthesis can be further subjected to technology independent and dependent optimizations [46,

BACKGROUND   33

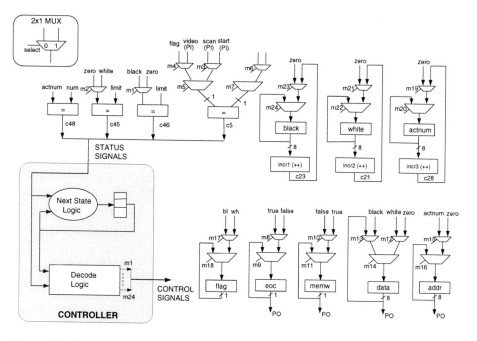

Figure 2.7. Structural RTL implementation of the barcode pre-processor example

61, 62] before passing the optimized RTL circuit on to logic and layout synthesis tools. In order to drive such optimizations, RTL estimation tools for area, delay, and power have been developed [1, 63, 64, 65, 66, 67].

2.4 HIGH-LEVEL SYNTHESIS APPLICATION DOMAINS

The evolution of high-level synthesis techniques has been driven by specific application domains, such as the data-flow or arithmetic intensive domain, that includes digital signal and image processing, graphics, and several multimedia applications, and the control-flow or decision intensive application domain, that includes networking/telecommunication protocols, controllers, *etc.* Each application domain has several distinct characteristics that need to be considered while developing efficient high-level synthesis techniques. Most high-level synthesis tools are well-suited to only one application domain. However, a large number of designs, in practice, contain a significant mix of control and data flow. Moreover, the tool development and maintenance costs associated with having significantly different synthesis algorithms for each domain are high. Hence, it is desirable for high-level synthesis tools to be able to handle designs that belong to any of the application domains as well as mixed designs.

Table 2.1 presents some distinguishing characteristics of data-flow intensive and control-flow intensive designs. The behavioral descriptions of data-flow intensive designs are dominated by arithmetic operations such as addition, subtraction, and multiplication, while those of control-flow intensive designs are dominated by nested conditional constructs, data-dependent loops, and comparisons, with very few arithmetic operations. The iteration constructs present in data-flow intensive designs are typically static, *i.e.* loops are either executed infinitely (as in the case of digital filters that operate on an incoming stream of data), or executed a fixed number of times independent of the input data. On the other hand, data-dependent loops are abundant in control-flow intensive designs, and it is difficult to statically predict the number of times a loop body is executed. As a result, the performance (number of clock cycles required to compute the output) of control-flow intensive designs is highly dependent on the input data. Scheduling techniques for data-flow intensive designs focus on extracting and exploiting the inherent parallelism or concurrency in the algorithm, while scheduling techniques for control-flow intensive designs must consider mutual exclusion of operations in the algorithm that arises due to the abundance of conditional constructs. Mutually exclusive operations on different branches of a conditional construct can never be simultaneously executed. The area, delay, and power of structural RTL implementations of data-flow intensive designs are dominated by arithmetic units and registers in the data path, while in the case of control-flow intensive designs, they are dominated by non-arithmetic

Table 2.1. Characteristics of data-flow and control-flow intensive applications

LEVEL OF ABSTRACTION	DATA-FLOW INTENSIVE	CONTROL-FLOW INTENSIVE
BEHAVIORAL	Data flow dominates	Control flow dominates
	Arithmetic operations	Data transfer, comparisons, bit-vector manipulation
	Data-independent loops, iterations known at compile time	Data-dependent loops, iterations determined dynamically
FUNCTIONAL RTL (Scheduled)	Parallelism	Mutual exclusion through nested conditionals
	Critical path simple, known statically	Critical path involves data-dependent loops
STRUCTURAL RTL	Dominated by arithmetic units (multipliers, adders, subtracters)	Dominated by multiplexers, comparators, counters, bit-manipulation units

units like multiplexers, bit-manipulation units, and comparators. The controllers required for data-flow intensive designs are simple and often no more than counters. As a result, the controller has very little impact on the area, delay, and power of the circuit. For control-flow intensive designs, on the other hand, even when the controller itself does not account for a large portion of the total delay and power, it can significantly affect the total circuit delay and power due to the composition of long paths through the control logic, and the effect of glitching activity at control signals. While most research on high-level synthesis techniques has targeted the data-flow intensive application domain, many designs, in practice, have significant control flow as well. Subsequent chapters present power analysis and optimization techniques for both data-flow intensive and control-flow intensive designs.

3 ARCHITECTURE-LEVEL POWER ESTIMATION

The register-transfer or architecture level is the design entry point for most designs today. Power estimation at this level of the design hierarchy is extremely important in order to (i) verify that power budgets are roughly met by the different parts of the design and the entire design, and (ii) evaluate the effect of various high-level optimizations, which have been shown to have a much more significant impact on power than lower-level optimizations. Architecture-level power estimation tools typically trade off some amount of accuracy for a drastic improvement in efficiency compared to low-level power estimation tools. The improved efficiency is due to the elimination of the need to obtain a gate- or transistor-level netlist, and the reduced complexity of analysis of RTL designs as compared to lower-level netlists. RTL design descriptions include various macroblocks like ALUs, vector logic operators, memories, register files, multiplexers, *etc.*, (which may be instantiated from a component library), as well as some amount of random or control logic, which may often be described functionally (*i.e.* without complete information about structure). This chapter describes the techniques that are used in architecture-level power estimation tools, including analytical power models, empirical activity and power macromodeling, sampling-based estimation, and models for control logic.

3.1 ANALYTICAL POWER MODELS

Analytical power modeling techniques attempt to correlate power consumption to measures of design complexity. Such techniques typically require very little information about the actual implementations of the macroblocks and random logic. For example, the Chip Estimation System [68] uses the following expression to compute the average power consumed by a macroblock:

$$P = GE \cdot \left(E_{typ} + C_L \cdot V_{dd}^2 \right) \cdot f \cdot A_{int} \qquad (3.1)$$

where GE is the estimated gate count of the block in terms of a basic gate type (e.g. two-input NAND), E_{typ} is the average energy consumed by an instance of the basic gate type when its output is switching, C_L is the estimated average load capacitance per gate, f is the clock frequency, and A_{int} is the estimated average switching activity factor. This approach is applicable to logic blocks designed using cell-based technologies, and is not very accurate for some parts of a chip such as the clock network, I/O, and memory blocks, as well as custom-designed macroblocks.

In [69], distinct modeling techniques were proposed for different parts of a chip, such as memory, clock, logic, interconnect, and circuits that drive off-chip loads. The power consumed in memories can be divided into four components – the cell array, the row decode, the column selection logic, and the read/write circuitry. The power consumed in a (static) six transistor memory cell array that consists of 2^n cells organized into 2^{n-k} rows and 2^k columns is given by:

$$P_{memcell} = \frac{2^k}{2} \cdot \left(c_{int} \cdot l_{column} + 2^{n-k} \cdot C_{tr} \right) \cdot V_{dd} \cdot V_{swing} \cdot f \qquad (3.2)$$

where c_{int} is the wiring capacitance per unit length, l_{column} is the bit-line (column) length, C_{tr} is the transistor drain capacitance, and V_{swing} is the bit-line swing. The above power represents the charging and discharging of the interconnect and transistor drain capacitances that are associated with the $bit/\overline{bit}$ lines, which typically forms the dominant part of the memory power. It is interesting to note that the above component is independent of data statistics, since both the bit and $\overline{bit}$ lines are pre-charged, and exactly one of them is discharged independent of whether the value stored in the memory cell is 0 or 1.

The clock power can be calculated from an architectural floorplan, by assuming a particular clock distribution network topology (*e.g.* H-tree), and sizing schemes for the clock buffer and the various branches of the clock tree.

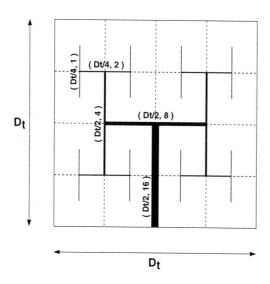

Figure 3.1. Estimating clock line capacitance

Example 3.1 *Consider the chip floorplan and clock tree shown in Figure 3.1. The chip is assumed to be of dimension $D_t \times D_t$. Suppose that the width of the leaves of the clock tree is the minimum wire width w_{min}, and that the width of the wires is reduced by a factor of two at each branching point. One segment at each level of the clock tree in Figure 3.1 is annotated with its length and width in terms of the chip dimension and minimum wire width, respectively (the dimensions of all segments at the same level are identical). The capacitance of the clock distribution* network is then given by:

$$C_{clkwire} = \left(16.\frac{D_t}{2} + 8.\frac{D_t}{2} + 2.4.\frac{D_t}{2} + 4.2.\frac{D_t}{4} + 8.1.\frac{D_t}{4}\right).c_{int} = 20.D_t.c_{int} \quad (3.3)$$

where c_{int} is the capacitance of a minimum-width wire per unit length.

The clock power consumption can be calculated as follows.

$$P_{clock} = \frac{1}{2}.C_{totalclock}.V_{dd}^2.f.2 = C_{totalclock}.V_{dd}^2.f \quad (3.4)$$

The term $C_{totalclock}$ in the above equation refers to the equivalent total capacitance that is associated with the clocking network, and can be computed using the equation:

$$C_{totalclock} = (1 + k_{driver}).(C_{clkwire} + C_{clkload}) \quad (3.5)$$

where k_{driver} is the clock driver ratio, which means that the clock driver capacitance is k_{driver} times the total capacitive loads associated with the clock network. The value of k_{driver} varies depending on the performance requirements of the system and the buffering scheme (*e.g.* distributed buffers, lumped buffer, *etc.*) used. A typical number for a conventional system is 0.3 [69]. $C_{clkwire}$ is the wiring capacitance associated with the clock distribution network, and can be estimated as described in Example 3.1. $C_{clkload}$ is the total load capacitance driven by the clock network, and can be estimated using the number of latches or flip-flops in the design, and the load presented by a typical latch or flip-flop to its clock input.

Information-theoretic approaches

The analytical power estimation techniques presented above either assumed a fixed activity factor, or relied on the user to provide an average switching activity factor. Some analytical techniques have been developed to estimate average activity factors for logic blocks based on measures of complexity of the functions they compute, and the signal statistics at the block inputs and outputs [70, 71]. In [71], power consumption in a macroblock is approximated as the product of its aggregate physical capacitance and the average switching activity of all its nodes. Consider a logic block that consists of N gates. Ignoring the short-circuit and leakage components, the average power dissipation can be written as

$$P \propto \sum_{i=1}^{N} C_i.D_i \approx C_{agg}.D_{avg} \qquad (3.6)$$

where C_i and D_i are the equivalent output capacitance and output switching activity for gate i, D_{avg} represents the average switching activity or transition density at the outputs of the gates in the block, and C_{agg} represents its aggregate physical capacitance. The area of a logic block can be used as an approximate measure of its aggregate physical capacitance. The area of a circuit can itself be estimated, from its functional description, using the total entropy at its outputs, H_O, when random sequences are applied at its inputs [72, 73].

$$\begin{aligned} Area &\propto 2^n H_O \; for \; n \leq 10 \\ &\propto \frac{2^n}{n}.H_O \; as \; n \to \infty \end{aligned} \qquad (3.7)$$

The above model may yield significant overestimates in area for large values of n due to the exponential dependence on n. A more accurate high-level area estimation procedure was presented in [74], based on the use of a metric called *average cube complexity*, which is the average literal count of the prime implicants of the function, in addition to entropy. Since enumeration of the prime implicants of a function can be highly computationally intensive, a Monte Carlo procedure can be used to estimate the average cube complexity.

The average switching activity at the outputs of the gates in a logic block can be approximated using the average entropy of the outputs of all the gates [71], which can be estimated from the entropies at primary inputs and primary outputs using the following equation:

$$D_{avg} \approx H_{avg} \approx \frac{2/3}{N_I + N_O}.(H_I + 2H_O) \qquad (3.8)$$

where N_I and N_O represent the number of primary inputs and outputs, and H_I and H_O represent the input and output entropies, respectively. The above equation is based on the assumptions that the entropy of all the gates at a given topological level in the circuit decreases quadratically with the level, and that the average number of gates at a level is $\frac{N_I+N_O}{2}$.

The input and output entropies of each logic block in an RTL circuit can be calculated through architectural simulation. However, in some situations, *e.g.* when rapidly evaluating several candidate architectures, simulation at each step may not always be possible. In such situations, propagation techniques for average entropy can be used [70]. For example, for data path blocks, the input entropies can be propagated to the outputs using the notion of *information transmission coefficient* (ITC). The ITC at the output of a logic block is computed using the ITC values at the block inputs using the following equation:

$$ITC_{out} = ITC_{block} \cdot \sum_{i=1}^{N_I} \frac{w_i}{W} \cdot ITC_{in} \qquad (3.9)$$

where N_I is the number of inputs to the logic block (some of which may be multi-bit signals), w_i is the bit-width of input i, $W = \sum_{i=1}^{N_I} w_i$, and ITC_{block} is a constant that depends on the functionality of the block. ITC_{block} can be pre-derived through a characterization process for typical RTL blocks. The above propagation technique is not well-suited to some parts of the circuit such as control logic. In such situations, the output entropies may be computed using a scaling factor based approach, as follows.

$$H_O = \frac{H_I}{f_{eff}^{N/2}} \qquad (3.10)$$

where f_{eff} is a scaling factor that can be computed from a gate-level netlist of the logic block, using the number of gates of each type or function (*e.g.* NAND, NOR, XOR, *etc.*), and the entropy scaling factor of each gate (*e.g.* XOR is entropy preserving, NAND is entropy decreasing by a factor of $\sqrt{2}$, *etc.*).

3.2 CHARACTERIZATION BASED ACTIVITY AND POWER MACROMODELS

A popular approach for estimating power in architectural building blocks is to construct a *macromodel* by obtaining and characterizing a lower-level implementation (that may be already available in the case of "hard" or "firm" macroblocks, or may need to be synthesized in the case of "soft" macroblocks [1]). A lower-level estimation tool is used to perform several experiments to estimate the power consumption of the macroblock for various input sequences, called *training sequences*. Based on the power consumption characteristics of the macroblock for the training sequences, a macromodel or function is constructed that describes the power consumption of the block as a function of various parameters, *e.g.* the signal statistics of the block inputs and outputs. A similar procedure may be used to characterize the glitching activity at the output of a macroblock. Such a characterization needs to be done only once, and the information may be re-used when the same macroblock is encountered in the context of other designs. A typical high-level power estimation flow that uses macromodeling is shown in Figure 3.2.

Characterization-based macromodels are best suited for bottom-up and meet-in-the-middle design methodologies [44], where hard or firm macroblocks can be instantiated from a component library. The characterization process, which is used to construct the power macromodels, can be thought of as part of the library development process, which is an effort that can potentially be re-used for several designs. The accuracy of characterization-based macromodels stems from the fact that a lower-level implementation is used to construct the macromodels. However, the training sequences used to construct the macromodel cannot be exhaustive due to efficiency considerations. Hence, a macromodel is in some sense "biased" by the

---

[1] The hardness of a macroblock reflects the extent to which its implementation has been targeted to a particular technology library and process. Hard macroblocks are mapped to a specific technology library and laid out, while soft macroblocks typically refer to generic, unsynthesized HDL. Firm macroblocks may be optimized for a generic library, or mapped to a specific technology library but not laid out.

44   HIGH-LEVEL POWER ANALYSIS AND OPTIMIZATION

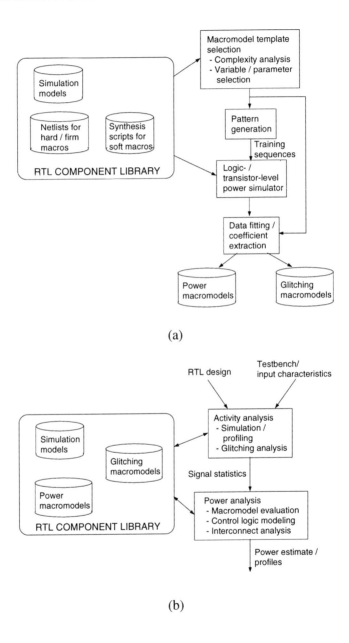

Figure 3.2. High-level power estimation flow using power macromodeling: (a) macromodel construction, and (b) power estimation

training sequences used during the characterization process. Since characterization is done upfront with possibly little information about the environment in which the components are going to be used, the training sequences may not very well represent real-life input sequences. In addition to the above problem, macromodels may introduce inaccuracies since the results of the characterization experiments are fit into a pre-determined function template or model, resulting in some errors due to interpolation or extrapolation. The characterization-based approach may also be used for soft macroblocks, by synthesizing the macroblock and using the gate- or transistor-level netlist for characterization, with the additional restriction that the synthesis results for the macroblock may not be the same when it is embedded within a larger design and the synthesis options/scripts used may differ.

In the *power factor approximation* (PFA) technique [75], the power consumption of a given type of functional block that is implemented using a given design style and operates at frequency f is estimated using the following equation.

$$P = \mathcal{K}.G.f \qquad (3.11)$$

where $\mathcal{K}$ is the PFA constant, and G is a measure of the hardware complexity of the functional block (that may be a function of parameters such as bit-width, *etc*). The PFA constant may be generated by characterizing one or a few implementations of the functional block. For example, assuming parallel multiplication, the hardware complexity G_{mult} of a multiplier is N^2 where N is the bit-width of the input operands. The PFA constant PFA$_{mult}$ for a multiplier was estimated in [75], based on published results for several fabricated designs, to be between 10 and 20 $\frac{femto\ Watt}{\#bit^2.Hz}$, depending on the technology. Similar models can be derived for the other components of a chip, including the I/O buffers and memories.

3.2.1 Activity-sensitive power macromodeling

The drawback of techniques such as PFA is that they do not account for the variation of power dissipation in different instances of a macroblock due to varying input signal statistics. In other words, they assume a fixed activity factor for the block, which may be very different from the activity factor for an actual instance. This can lead to very large estimation errors, as pointed out in [76]. Activity-sensitive power models [1, 2, 67, 76, 77, 78, 79] alleviate the above deficiency by constructing and utilizing a model for power consumption that is a function of the signal statistics at a macroblock's boundaries. Given a design that contains instantiations of macroblocks, a typical approach is to use architectural simulation to determine the signal statistics at the various macroblock boundaries, and feed the signal statistics for each macroblock instance to the corresponding power model to compute the power consumed in that instance.

For example, in the power estimation tool ESP [78], the power consumption in the data path components is modeled as consisting of a constant component P_{const} that is incurred whenever the macroblock is activated during the architectural simulation, and an activity-dependent component that is the product of the number of input bit transitions and a coefficient P_{change} that represents the power consumed per input bit transition. Transitions on the output bits of a macroblock can also be included along with input bit transitions in the model. Thus, the power P is given by

$$P = P_{const} + n * P_{change} \tag{3.12}$$

where n is the number of input/output bit transitions.

A more comprehensive activity-sensitive power analysis methodology was presented in the tool SPA [1, 76, 77]. Data path signals are modeled using the *dual bit type* (DBT) model, which is based on observations of the bit-level transition activity in typical fixed-point two's complement data streams. Figure 3.3 shows the bit-level transition activity ($Prob(0 \rightarrow 1)$) at the various bits of a multi-bit signal that represents a two's complement fixed-point number, for various sequences that have different word-level temporal correlations (ρ). The figure illustrates that the bits can be divided into two distinct regions.

- The least significant bits, or uniform white noise (UWN) bits, that tend to be random in nature.

- The sign bits whose activity depends heavily on the sign transition probability and, hence, on the (word-level) temporal correlation of the sequence of values appearing at the signal.

In addition, there is a transition region that lies in between the sign bits and the UWN bits. The breakpoints BP_0 and BP_1 that define the UWN and sign regions of a multi-bit signal can be calculated from its word-level signal statistics (mean μ, standard deviation σ and temporal correlation ρ) using the following equations.

$$\begin{aligned} BP_0 &= \log_2 \sigma + \log_2 \left(\sqrt{1 - \rho^2} + \frac{|\rho|}{8} \right) \\ BP_1 &= \log_2(|\mu| + 3.\sigma) \end{aligned} \quad (3.13)$$

Based on the DBT model for data path signals, black-box models can be constructed for macroblocks to estimate the capacitance switched in the macroblock for any desired input data statistics. The switched capacitance or power consumed

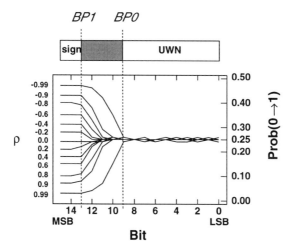

Figure 3.3. Relationship between word-level temporal correlation and bit-level transition activity [1]

48 HIGH-LEVEL POWER ANALYSIS AND OPTIMIZATION

in a macroblock can be thought of as a function of its complexity, as well as the switching activity within the macroblock. The complexity of a macroblock is modeled through an expression that relates the total equivalent capacitance in the block to parameters that may vary from module to module. The designer's insight and interaction are required in order to come up with accurate complexity models. For example, the total capacitance of an N-bit ripple-carry subtracter can be modeled using the following equation.

$$C_T = C_{eff}.N \qquad (3.14)$$

C_{eff} in the above equation refers to the effective switched capacitance per bit. Having a constant scalar value of C_{eff} for all bits of the subtracter will result in ignoring the effect of input statistics. Activity dependence can be added by treating different bits of the subtracter differently, depending on the data statistics at the individual input/output bits. This results in a multitude of capacitive coefficient values, rather than a single scalar value for C_{eff}.

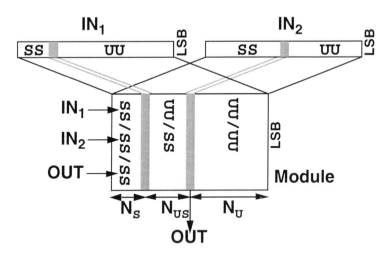

Figure 3.4. Transition template for a 2-input subtracter with misaligned breakpoints [1]

ARCHITECTURE-LEVEL POWER ESTIMATION   49

Table 3.1.   Capacitance coefficients for a 2-input subtracter [1]

	Transition templates	Capacitive coefficients			
	UU/UU	$C_{UU/UU}$			
	SS/SS/SS	$C_{++/++/++}$	$C_{++/++/+-}$	$C_{++/++/-+}$	$C_{++/++/--}$
		$C_{++/+-/++}$	$C_{++/+-/+-}$	$C_{++/+-/-+}$	$C_{++/+-/--}$
		...	...	...	...
		$C_{--/-+/++}$	$C_{--/-+/+-}$	$C_{--/-+/-+}$	$C_{--/-+/--}$
		$C_{--/--/++}$	$C_{--/--/+-}$	$C_{--/--/-+}$	$C_{--/--/--}$
Misaligned breakpoints only	UU/SS	$C_{UU/++}$	$C_{UU/+-}$	$C_{UU/-+}$	$C_{UU/--}$
	SS/UU	$C_{++/UU}$	$C_{+-/UU}$	$C_{-+/UU}$	$C_{--/UU}$

Assuming the DBT model, each of the two inputs of the subtracter can be divided into two bit-regions, the UWN bits and the sign bits. This leads to various possible "operating regions" for the subtracter bit-slices, as shown in the transition template of Figure 3.4. The subtracter bit-slices where both the operand bits belong to the UWN region are said to be operating in the UU/UU region [2], and have an associated capacitance coefficient $C_{UU/UU}$ as shown in Table 3.1. Next, consider the bit-slices where both the input bits correspond to sign bits. In this region, the capacitance switched per bit can significantly depend not only on the values of the input sign bits but also on the values of the output bit, which cannot always be determined solely based on the input sign bits (e.g. the subtraction of two positive numbers may result in a positive number or a negative number). Hence, such bit-slices are said to be operating in the SS/SS/SS region. Distinct capacitive coefficients are derived for this region for all possible combinations of pairs of sign bit values at both the input bits as well as the output bit (e.g. ++/++/+- represents two consecutive cycles such that in the first cycle one positive number is subtracted from another and the result is positive, and in the second cycle the two inputs are

---

[2] UU is used to indicate the transition activity, i.e. the previous and present values, at the input bits.

positive but the result is negative). In addition to the above regions, we may also have regions where one of the input bits is a sign bit and the other is a UWN bit (UU/SS and SS/UU). This happens when the breakpoints BP_0 and BP_1 of the two inputs to the subtracter are not aligned. Again, separate capacitive coefficients are calculated for distinct value pairs at the sign bits, leading to four coefficients each for the UU/SS and SS/UU regions. In all, there are a total of 73 capacitive coefficients for the ripple-carry subtracter. The number of capacitive coefficients becomes larger as the number of inputs to a module increases, and for multi-function units. The table of capacitive coefficients is constructed by generating specific pseudo-random pattern sequences that exercise all possible sign bit and UWN transitions, simulating a low-level implementation of the macroblock using the generated input sequences, and extracting best-fit capacitance coefficient values from the simulation data.

While the DBT-based modeling methodology accounts for the dependence of power consumed in a macroblock on the functional or zero-delay signal statistics at its inputs, it ignores the effect of glitching activity at a macroblock's inputs on its power dissipation [3].

3.2.2 Accounting for glitching power consumption

Glitch generation and propagation through circuit components may lead to the presence of significant glitching at data path as well as control signals. Ignoring glitching power consumption may lead to loss of absolute as well as relative accuracy in power estimation [67]. One approach to account for glitching power consumption during power macromodeling involves (i) building macromodels to estimate glitching activity at the outputs of various data path components (in the presence of glitching at their inputs), and (ii) building power macromodels that

---

[3] The functional or zero-delay statistics at signals in a circuit can be derived by assuming that all the circuit elements compute their results instantaneously. Transitions in a circuit that do not correspond to functional or zero-delay transitions are called glitches. Glitches do not contribute to the functionality of a circuit, and, hence, cause unnecessary power consumption.

ARCHITECTURE-LEVEL POWER ESTIMATION 51

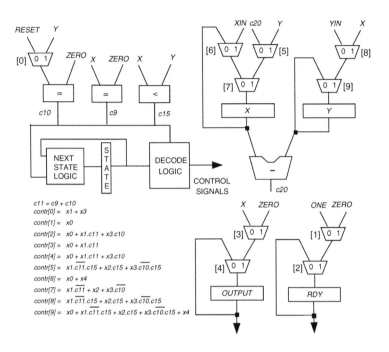

Figure 3.5. The GCD RTL circuit

depend on the glitching activity at the macroblock inputs in addition to other signal statistics [67]. The following example motivates the need to account for glitching power during high-level power estimation.

Example 3.2 *Consider the RTL circuit shown in Figure 3.5, that computes the greatest common divisor (GCD) of two numbers. The RTL blocks used in the GCD data path are one subtracter, three comparators (one less-than ($<$) and two equal-to ($=$)), registers, and multiplexers. The controller is sub-divided into the state register, the next state logic, and the decode logic that generates the control signals for the data path. The control expressions implemented by the decode logic are also given in Figure 3.5. Control signal $contr[i]$ feeds the select input of the multiplexer marked $[i]$. Similarly, data path signal $dp[i]$ corresponds to the output*

52   HIGH-LEVEL POWER ANALYSIS AND OPTIMIZATION

of the multiplexer marked [i]. *The variables* $x0, ..., x4$ *in the control expressions represent the decoded state variables. The control expressions also involve status signals generated from the data path like the outputs of comparators. While the data path typically consists of several pre-designed macroblocks, the control logic is often subject to logic synthesis optimizations before it is mapped to the technology library. Activity profiles for some of the data path signals of the* GCD *circuit, obtained by synthesizing and mapping it to NEC's CMOS6 library [80], and using a simulation-based power estimation tool [81] are shown in Figure 3.6.*

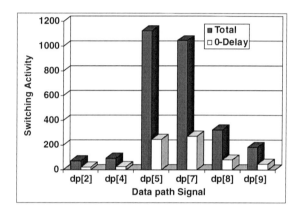

Figure 3.6.   Activity profiles for data path signals in the GCD circuit

The activity profiles show the total switching activities (including glitching) and zero-delay switching activities (excluding glitching) for selected data path signals in the circuit. The significant differences between the total and zero-delay switching activities indicate the importance of considering the effect of glitching activity in high-level power estimation.

In order to explore the ramifications of the inaccuracies in switching activity on macroblock power estimates, the following experiments were performed. First, the entire GCD *circuit implementation (gate-level technology-mapped netlist) was simulated using several typical input sequences, to estimate its average power* consumption. The power consumption was reported to be $1.64mW$. *This figure includes the effect of glitch propagation across RTL blocks, since the entire circuit was used in the simulation.* Next, an RTL simulation was performed using the same input sequences, and traces were collected for the inputs of each embedded circuit block. The RTL simulation results in zero-delay traces (that do not include glitches) at various data path and control signals. The implementation of each RTL block in the GCD circuit was simulated separately (the controller was considered as a single block) using the zero-delay traces derived in the previous step, and the individual power estimates were summed up to yield a power estimate of $1.32mW$ for the entire circuit (an under-estimate of 19.5%). While the above error represents an average over all the macroblocks in the GCD circuit, the errors in power estimates for individual macroblocks are even higher, often as high as 50%.

Word-level glitch generation and propagation models

A word-level *glitching activity macromodel* relates the glitching activity at the output of an embedded macroblock to zero-delay statistics at the input signals (*e.g.* mean, standard deviation, spatial and temporal correlations in the case of signals with numeric values), and glitching activity at the inputs themselves [67]. Glitch models can be constructed for various RTL library components through a process of characterization shown in Figure 3.2(a). The characterization process consists of constructing controlled experiments (simulation runs) by selectively varying one or more of the controllable variables (input zero-delay statistics and glitching activities), and observing the value of the dependent variable (glitching activity at the block output). The glitching activity function can be very complex and, hence, may not fit well into a simple expression. In such situations, the use of one or more *piecewise linear models* (based on lookup tables) is a possible macromodel template that is flexible and suited to automation.

Consider, for example, an 8-bit subtracter with inputs A and B, and output OUT. In general, the glitching activity at OUT can be written as follows.

$$Gl_{OUT} = f_{gl}(Mean_A, Mean_B, SD_A, SD_B, TC_A, TC_B, SC_{A,B}, Gl_A, Gl_B) \qquad (3.15)$$

The first seven parameters of $f_{gl}()$ represent the zero-delay signal statistics at A and B. $Mean_A$ represents the mean or expected word-level value represented by signal A, SD_A represents its standard deviation, TC_A is the temporal correlation coefficient that represents the correlation between consecutive values that appear at signal A, $SC_{A,B}$ is the spatial correlation coefficient of A and B. Gl_A represents the glitching activity at A. Parameters with subscripts B have a similar meaning. The brute-force approach for building a model for $f_{gl}()$ would involve discretizing the range of variation of each of the parameters with a desired granularity, generating input sequences that correspond to each possible set of values for the parameters, and simulating the implementation of the subtracter to observe the glitching activity at the subtracter's output for each case. Assuming that each parameter can assume k possible values, the above approach will require k^n simulations, where n is the number of parameters or independent variables considered. In the case of the subtracter, $n = 9$, and even assuming $k = 5$ leads to 1.95 million simulation runs! Clearly, the brute-force approach is undesirable, in spite of the fact that building the models is a one-time cost for a given component library.

Two techniques may be used to avoid the combinatorial explosion in the number of simulation runs required. The first technique, called *variable elimination*, attempts to reduce the number of independent variables in the glitching activity model by identifying those variables whose variations affect the dependent variable (output glitches) minimally. Techniques from multi-variable data analysis can be used for this purpose. Given a set of samples (each sample consists of a set of values for the independent variables $x_1, \ldots, x_n$, and the corresponding observed value that the dependent variable y assumes), the *ANOVA test* [4] is used to check whether the null hypothesis for any given variable x_i is true, *i.e.* whether different values of x_i had any impact on the observed sample values of y [82]. The second technique,

---
[4] ANOVA (Analysis Of VAriance) is a popular technique used for statistical inference and testing.

called *model decomposition*, attempts to decompose the function $f_{gl}()$ into multiple sub-functions by partitioning the set of parameters into smaller groups of variables such that the effects of variables from different groups on the dependent variable interact minimally. Again, it is possible to use standard ANOVA techniques to obtain a quantitative evaluation of the interaction of the effects of two independent variables on the dependent variable from a given set of samples.

For example, in the case of the subtracter, the basic model of Equation (3.15) can be decomposed into the following equation.

$$Gl_{OUT} = f_{gl_1}(Mean_A, Mean_B) * f_{gl_2}(SD_A, SD_B) * f_{gl_3}(TC_A, TC_B) * \\ f_{gl_4}(SC_{A,B}) * f_{gl_5}(Gl_A, Gl_B) \qquad (3.16)$$

The independent variables have been partitioned in the above equation into the groups, $\{Mean_A, Mean_B\}$, $\{SD_A, SD_B\}$, $\{TC_A, TC_B\}$, $\{SC_{A,B}\}$, and $\{Gl_A, Gl_B\}$. The above partition was based on the observation that variables within each group have a significant interaction in their effect on the dependent variable, while the variables in distinct groups are reasonably independent in their effects. As before, assuming that the domain for each parameter is discretized into five distinct regions, it is necessary to perform simulations for $5^2 + 5^2 + 5^2 + 5^1 + 5^2 = 105$ different sets of parameter values, which can be performed much more efficiently compared to the approach of building a single huge piecewise linear model from Equation (3.15).

The piecewise linear models that constitute the glitching activity macromodel for the 8-bit subtracter are shown in Figure 3.7. Figure 3.7(a) models the glitching activity at the subtracter output as a function of the means of the sequences at the two inputs. Figures 3.7(b)-3.7(e) represent multiplicative correction factors that account for the effects of the remaining variables. Note that in order to generate the model for $f_{gl_5}()$, it is necessary to generate input sequences to the subtracter

56   HIGH-LEVEL POWER ANALYSIS AND OPTIMIZATION

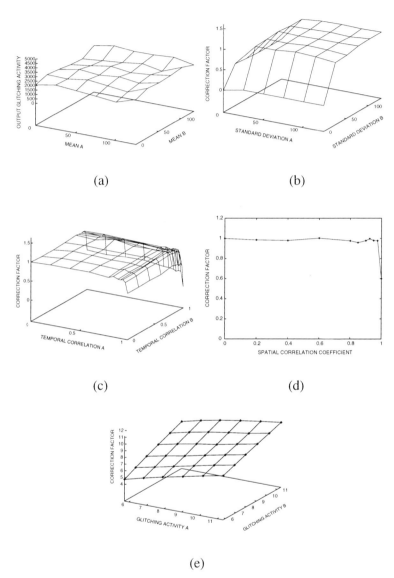

Figure 3.7.   Glitching activity models for an 8-bit subtracter

that have varying glitching activities, in addition to having the required zero-delay statistics.

Bit-level glitch generation and propagation models

Bit-level modeling allows us to build more accurate glitching activity models, and to consider the effects of bit-level statistics that may not be well-reflected by word-level signal statistics in certain situations. Let us first consider the generation of glitches in a multiplexer when its inputs are glitch-free. Suppose a multiplexer bit-slice has two data input bits A_i and B_i, and one select input Sel. The number of distinct input vector pairs that can be applied at A_i, B_i, and Sel is $2^3 * 2^3 = 64$. Since the above number is small, it is possible to simulate the implementation of a one-bit multiplexer for the exhaustive set of 64 input vector pairs, and build a look-up table, $Mux\_gl\_gen[]$, that stores the glitches generated at the output for each vector pair. For a given input vector pair, glitch generation at each bit-slice of a multiplexer is estimated by looking up the appropriate entry of the $Mux\_gl\_gen[]$ table.

The output of a multiplexer can also be glitchy due to the propagation of glitches from the data and select inputs. The propagation of glitches from a data input to the multiplexer output can be modeled as being "regulated" by the probability that the glitchy data input is selected. For a one-bit slice, assuming A_i is selected when $Sel = 0$, the glitching activity at the multiplexer output due to propagation from A_i is given by $Gl(A_i) * P(Sel = 0)$. A similar explanation holds for the propagation of glitches from B_i. The propagation of glitches from the select signal of a multiplexer is affected by the spatial correlation between the data inputs in addition to the signal probabilities [67]. The model for glitch propagation from the select signal of the multiplexer to its output is given by the following equation.

$$Gl(Sel) * (D_{01} * P(A_i = 0, B_i = 1) \ + \ D_{10} * P(A_i = 1, B_i = 0)$$
$$+ \ D_{11} * P(A_i = 1, B_i = 1))$$

58   HIGH-LEVEL POWER ANALYSIS AND OPTIMIZATION

The probabilities $P(A_i = 0, B_i = 1)$, $P(A_i = 1, B_i = 0)$, and $P(A_i = 1, B_i = 1)$ are monitored for each multiplexer bit-slice during zero-delay simulation or analysis. The constants D_{01}, D_{10}, and D_{11} depend on the exact implementation of the multiplexer, and may be computed by performing experiments using a circuit configuration such as the one shown in Figure 3.8. The comparator is used

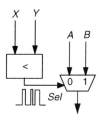

Figure 3.8.   Circuit used to compute the coefficients D_{01}, D_{10}, and D_{11}

to generate glitches at the select input to the multiplexer by feeding appropriate vector sequences at its inputs. In order to calculate D_{01}, the multiplexer's data inputs are fixed to $A = 0...0$ and $B = 1...1$. The glitching activity, $Gl(OUT_i)$, at the output of the ith bit-slice of the multiplexer is recorded for each i. Note that $Gl(OUT_i)$ also includes the effects of glitch generation in the multiplexer. Hence, zero-delay traces at the multiplexer inputs are used to estimate glitch generation in the multiplexer through the look-up table based procedure described earlier. The estimated glitch generation value is subtracted from $Gl(OUT_i)$. The resulting difference is averaged over all i, and divided by the value of $Gl(Sel)$ to obtain the value of D_{01}. The coefficients D_{10} and D_{11} are calculated similarly.

In summary, the glitching activity at the output of an n-bit multiplexer with data inputs A and B, select input Sel and output OUT is calculated using the following equations.

$$Gl(OUT) = \sum_{i=1}^{n}(Gl\_Gen(i) + Gl\_Prop\_From\_A_i + \qquad (3.17)$$
$$Gl\_Prop\_From\_B_i + Gl\_Prop\_From\_Sel)$$

$Gl\_Gen(i)$ is accrued during zero delay simulation by looking up the $Mux\_gl\_gen[]$ table.

$$Gl\_Prop\_From\_A_i = Gl(A_i) * P(Sel = 0)$$
$$Gl\_Prop\_From\_B_i = Gl(B_i) * P(Sel = 1)$$
$$Gl\_Prop\_From\_Sel = Gl(Sel) * (D_{01} * P(A_i = 0, B_i = 1) +$$
$$D_{10} * P(A_i = 1, B_i = 0) +$$
$$D_{11} * P(A_i = 1, B_i = 1))$$

3.2.3 Bit-level and cycle-accurate power macromodels

Bit-level modeling allows us to build more accurate power macromodels in some situations. For instance, if a bit-vector signal b consisting of n bits is actually a concatenation of two smaller bit-vectors b_1 and b_2 that consist of k and $n - k$ bits, respectively, it may be desirable to compute signal statistics for b_1 and b_2 separately. The extra computational effort spent here is well-justified for control-flow intensive designs where the total circuit power consumption may be dominated by power consumption in multiplexers, registers, and bit-manipulation operators.

Aggregate power macromodels compute the power consumed in a macroblock as a function of data statistics (*e.g.* mean, standard deviation, *etc.*) that are aggregated over a large number of clock cycles. They are either inapplicable or their accuracy tends to be very limited in predicting power on a cycle-by-cycle basis. Cycle-accurate power estimation is important when feedback, such as which clock cycles (or control steps) a macroblock consumes the most power in, is needed, or to plot the power consumption distribution of the entire circuit over time.

A popular bit-level cycle-accurate power macromodeling technique is the *peripheral capacitance model* (also referred to as the linear regression model), where

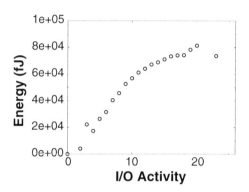

Figure 3.9. Variation of energy consumption with input/output switching activity for an 8-bit carry-lookahead adder [2]

the power consumption in a macroblock is modeled by computing a weighted average of the transition activities at its input and output bits [2, 83, 84]. This model is based on the observation that the variation of power consumed in a macroblock with the total switching activity at its inputs and outputs can often be well-approximated by a linear function, especially for arithmetic circuits. For example, Figure 3.9 shows a scatter plot of the power consumed in an 8-bit carry-lookahead adder with the total switching activity at its inputs and outputs, for several random input sequences. The fact that the points on the plot are close to a straight line suggests the use of a linear relationship to model power consumption. Consider a macroblock with m input bits $i_1, \ldots, i_m$ and n output bits $o_1, \ldots, o_n$. The peripheral capacitance model for the macroblock is given by the following equation.

$$P = \frac{1}{2} \cdot V_{dd}^2 \cdot f \cdot \left(\sum_{k=1}^{m} C_{i_k} \cdot A_{i_k} + \sum_{k=1}^{n} C_{o_k} \cdot A_{o_k} \right) \quad (3.18)$$

In the above equation, A_p is the switching activity at pin p, and C_p is the constant associated with the pin. C_p can be thought of as an equivalent capacitance attached to pin p. Hence, the above model approximates the power consumed in a macroblock as a set of equivalent capacitances at the boundaries, or periphery, of the block, as shown in Figure 3.10.

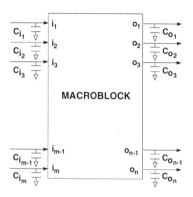

Figure 3.10. Peripheral capacitance model

The values of the various constants can be derived by simulating the implementation of the macroblock using a large set of input sequences, and fitting the data obtained (switching activity at each pin for each vector pair and the power consumed in the macroblock under the application of the vector pair) to a linear function. Using a linear function enables us to leverage off the well-developed theory of linear regression to construct the model from a set of simulation data points. However, in some situations, attempting to fit a highly non-linear function into a linear model may result in significant errors. In particular, for macroblocks with data inputs as well as control inputs (*e.g.* multi-function ALUs), while the linear model applies well to switching activity at data inputs, it is not well-suited to the control inputs which tend to have a much stronger influence on the power consumption in the macroblock. In such situations, one possible approach is to construct multiple linear power models, one for each distinct set of values assumed by the variables representing control inputs. In situations where the inputs to a block are not explicitly partitioned into control and data inputs, the variance of the power values with respect to the values at each data input can be computed from simulation data and used to identify inputs based on which the model should be decomposed [2].

While the peripheral capacitance model offers a simple and intuitive way to model the power consumed in a macroblock, its accuracy is limited due to the linear relationship that is assumed between power consumption and switching activity at the various pins. In reality, the power function may be a highly complex, non-linear function for some macroblocks, and assuming any fixed relationship or function template may be inaccurate. Recognizing this, a general macroblock modeling technique called *energy clustering* was presented in [85]. The power or energy consumed in a combinational macroblock is expressed as a function of the previous and present input vectors. An exhaustive "energy table" that relates each possible previous and present input vectors to the energy consumed by the circuit under their application would consist of 2^{2n} entries where n is the number of inputs to the circuit. For a sequential circuit, the initial state of the circuit before application of the vector pair should also be considered, requiring 2^{2n+s} entries where s is the number of storage elements (latches or flip-flops). Clearly, the size of exhaustive energy tables is prohibitively large even for moderate n and s. In energy clustering, the basic idea is to construct a compact representation for the energy truth table, by combining several entries with similar energy consumption values into single entries called *energy cubes*, very similar to the way in which minterms are combined into cubes in two-level circuit optimization in logic synthesis [47]. The energy estimate for an energy cube is the average energy of all the simulation points that it covers. The error in the energy estimate of an energy cube can be captured by computing the average, root mean-square, or maximum deviations (or a combination thereof) of the energy values for the individual points covered by the cube with respect to the cube's energy estimate. Automatic heuristic procedures to construct a clustering-based energy model for a macroblock are described in [85]. These procedures cluster the data points resulting from the simulation into energy cubes that cover the entire Boolean space, such that the energy estimate error of each cube is below a user-specified threshold.

3.2.4 Improving macromodel efficiency with statistical sampling

While power estimation using macromodels tends to be more efficient than gate-level power estimation, it may still be inefficient when invoked iteratively for the purpose of manual re-design or automatic synthesis. For example, the peripheral capacitance model requires, on a per-cycle basis, the computation of the switching activities for each input and output bit of the macroblock.

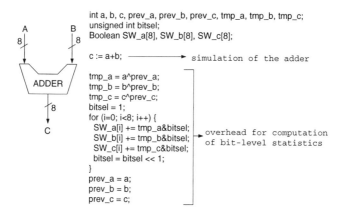

Figure 3.11. Overhead for computing bit-level statistics

Example 3.3 *Consider an adder that computes the sum of two 8-bit numbers. During RTL simulation, it is common to model the input operands of the adder as integer variables, and use the host computer's addition operation to simulate the execution of the adder. The extra operations required to compute the transition activity information at the inputs of the adder for each clock cycle are indicated in the pseudo-code shown in Figure 3.11. The extra computation required to calculate bit-level statistics is much larger than the computation required to just simulate the adder. As a result, the RTL simulation process may be slowed down significantly.*

The idea of statistical sampling has been used in the context of logic- and transistor-level power estimation [86]. Given user-specified confidence and error

64   HIGH-LEVEL POWER ANALYSIS AND OPTIMIZATION

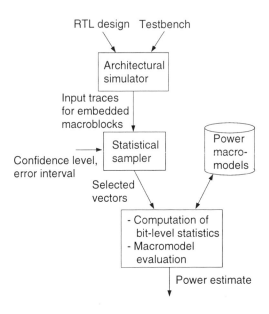

Figure 3.12.   Statistical sampling to improve the efficiency of high-level power estimation

levels, statistical criteria are used to determine the number of vectors that need to be simulated. The operation of a macromodeling based power estimator with statistical sampling is illustrated in Figure 3.12. The required sample size is computed from the user-specified confidence level and error intervals. The RTL simulator is used to simulate the RTL circuit under the entire input sequence. However, the computation of the signal statistics required to evaluate the power macromodel and the evaluation of the macromodel are performed only at selected clock cycles. If random sampling is used, the clock cycles are selected randomly among the set of all clock cycles simulated. More sophisticated sequence compaction strategies [87, 88] may also be used to obtain a reduced set of patterns for computation of signal statistics and macromodel evaluation. This idea may be applied to both aggregate and cycle-accurate power macromodels.

3.2.5 Improving estimation accuracy using adaptive macromodeling

The techniques described in the previous sections are based on static macromodels, *i.e.* the macromodels are constructed once for a given macroblock, and not changed from one design that uses the macroblock to another. A significant source of error in static macromodel-based power estimation is the bias, or the dependency of the macromodel on the set of training patterns used. Macromodels are most accurate when the training patterns reflect well the input sequences seen by the macroblock instance. However, the actual input sequences can vary significantly depending on the design in which the macroblock is used, and its environment.

Adaptive macromodeling addresses the above drawback by modifying or adapting the macromodel using selected vectors from traces for the macroblock's input/output signals that are derived from a simulation of the complete design being analyzed [84] (see Figure 3.13). A gate- or transistor-level power simulator is invoked for the selected vectors or vector sequences. Apart from providing accurate power estimates for the selected vectors, the results of the lower-level power simulator are used to correct or adapt the macromodel using regression analysis to improve the estimation accuracy for the remaining vectors as well.

In particular, let us consider the peripheral capacitance model for a macroblock, where power consumption is modeled as a function of the present and previous input vectors to the macroblock. Let S be the set of all input vector pairs for an embedded macroblock, derived from RTL simulation, with $|S| = N$. Suppose we select a set $s \subset S$ of input vectors pairs for gate-level power simulation, with cardinality $|s| = n$. Let x_i represent the power estimate computed using the (static) macromodel for vector pair i, $i \in \{1, \ldots, N\}$. Let y_i represent the power estimate obtained from the gate-level power simulator for vector pair i, $i \in \{1, \ldots, n\}$. The average power consumption, y_{avg}, for the entire input sequence can be computed using the following equations.

$$y_{avg} = \frac{1}{N} \cdot \left(\sum_{i \in s} y_i + \sum_{i \notin s} y_i \right)$$

66 HIGH-LEVEL POWER ANALYSIS AND OPTIMIZATION

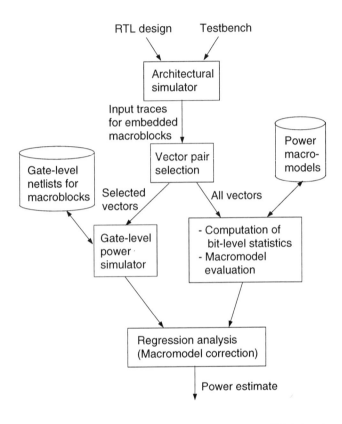

Figure 3.13. Adaptive macromodeling to improve the accuracy of high-level power estimation

$$\sum_{i \notin s} y_i = (N - n).\alpha + \beta \sum_{i \notin s} x_i \qquad (3.19)$$

The values of the coefficients α and β can be determined using least-square-error linear regression analysis using the available data (x_i, y_i), where $i \in s$.

3.3 POWER AND SWITCHING ACTIVITY ESTIMATION TECHNIQUES FOR CONTROL LOGIC

Architectural power estimation tools must include techniques for modeling not only the data path components or macroblocks, but also the control or random logic parts of a design. In addition to computing the power consumed in the control logic, it is also very important for architectural power estimation tools to take into account the impact of the signal statistics at control signals on the power consumption of the rest of the design. The zero-delay statistics at the control signals are important since values appearing at control signals may be highly correlated spatially as well as temporally [89]. In addition, it is also important to take into account the glitching activity at the control signals since it can have a significant impact on the power consumption in the rest of the design [67], as illustrated by the following example.

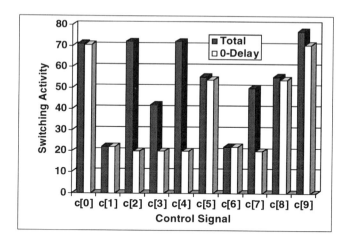

Figure 3.14. Activity profiles for control signals in the GCD circuit

Example 3.4 *Consider again the GCD RTL circuit shown in Figure 3.5. The power consumption of the gate-level technology-mapped implementation of the circuit*

was evaluated for a given testbench. The activity profiles for the control signals are shown in Figure 3.14 ($contr[i]$ is reduced to $c[i]$), and show the total and zero-delay switching activities at each control signal during the entire simulation run. The large difference between the total and zero-delay activities in some cases implies that control signals can have a lot of glitching activity. The impact of this glitching activity on the power consumption of the circuit was evaluated as follows. The implementation of the entire circuit under the given testbench was found to be $1.64mW$, as reported earlier. In order to evaluate the effect of glitching activity at control signals on the power consumption of the circuit, the circuit was partitioned into the controller and the data path. RTL simulation was used to gather traces for the interface between the controller and data path. Power estimation was performed separately for the implementation of the controller and data path, using their respective (zero-delay) input traces. The sum of the individual power estimates was found to be $1.45mW$, indicating that the glitching activity at the controller-data path interface has a significant impact on the total power consumption.

3.3.1 Controller power consumption

In the *activity-based control* (ABC) approach to control logic power modeling [90], the power consumed in the controller is estimated given the implementation style (*e.g.* ROM, PLA, standard cell, *etc.*) of the controller and its state machine description, say in the form of a state table. Target-independent complexity parameters that are used in the power model include:

- N_I - the number of inputs of the combinational logic of the controller (including the primary inputs as well as the present state lines),

- N_O - the number of outputs of the combinational logic (including primary outputs as well as next state lines),

- N_S - the bit-width of the state register, and

- N_M - the number of minterms in a minimized implementation of the controller.

N_I, N_O, and N_S can be estimated given an encoding style attribute such as minimal binary, one-hot, *etc.* N_M is estimated given a state assignment (or using a random state assignment if none is provided) by performing fast logic minimization, and using the number of minterms in the minimized implementation as an estimate of N_M. The model also uses activity parameters such as α_I, α_O, and α_S - the switching activities at the inputs to the combinational logic, outputs of the combinational logic, and the state lines, respectively. Activity parameters can be derived from an architectural simulation of the design. For example, the switched capacitance C^T for a standard-cell based controller is given by the following equations.

$$\begin{aligned} C^T &= C^{reg} + C^{CL} \\ C^{reg} &= C_0^{reg}.\alpha_S.N_S \\ C^{CL} &= C_0^{CL}.\alpha_I.N_I.N_M + C_1^{CL}.\alpha_O.N_O.N_M \end{aligned} \quad (3.20)$$

The capacitance coefficients C_0^{reg}, C_0^{CL}, and C_1^{CL}, depend on the library and the synthesis tools being used, and are measured through a characterization process. Controllers of varying complexities and displaying varying input and output activities are synthesized by generating random controller state tables and input patterns. A low-level implementation is simulated to obtain data points, that are then used to obtain best-fit values for the capacitive coefficients.

Another approach is to perform a fast synthesis of the control logic, and perform an architectural simulation of the design with the control logic behavior being replaced by its fast-synthesized netlist. Activities at each signal in the control logic can be monitored, and combined with average gate output capacitance values to yield an estimate of the control logic power consumption. The efficiency of architectural simulation is preserved if the control logic represents a relatively small portion of the complete design. Since very crude or no delay information is available during RTL simulation, however, the activity information for the control logic is inaccurate since it does not include effects such as glitching.

3.3.2 Estimating glitching activity in the control logic

This section presents techniques to estimate glitch generation and propagation through the controller, which can significantly affect the total power consumption as shown in the previous section. The controller's inputs are the status signals from the data path (typically outputs of comparators or combinations thereof), while its outputs are the control signals that feed the data path. Clearly, glitch generation and propagation in the control logic can be exactly estimated only if detailed information regarding the structure of the controller implementation and delays are provided. However, the final implementation of the controller is typically *not available* during high-level design iterations, since the control logic is subject to significant logic synthesis optimizations, and completely synthesizing the controller within each iteration, of say a high-level synthesis tool, is too computationally expensive. Estimation techniques for other design metrics, *e.g.* area and delay [65, 66], utilize the high-level representation used for the control logic to derive estimates that are efficient and reasonably accurate in practice. The switching activity at control signals is computed by combining zero-delay activity numbers derived from RTL simulation, with glitching activity estimates derived from *control expressions*, as explained next [67].

The control logic is typically represented as *control expressions* during the high-level synthesis process. These control expressions are usually expressed in the form

$$contr = \sum_i xi . \left(\prod_j cij \right) \quad (3.21)$$

where xi represents a decoded controller state variable (corresponding to controller state si), cij represents a status signal, which is typically the output (or inverted output) of a comparator from the data path, $\sum$ represents the Boolean OR operation and $\prod$ and . represent the Boolean AND operation. Each product term in the control

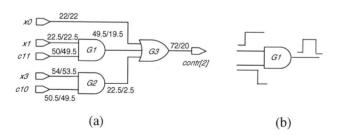

Figure 3.15. (a) Implementation of control signal $contr[2]$, and (b) generation of glitches at gate $G1$

expression flags the occurrence of a particular combination of values at the status signals when the controller state is s_i. The status signals (c_{ij}) may themselves carry glitches, that propagate through the control logic, causing the control signals to be glitchy. On the other hand, the control logic can also generate a significant amount of glitches, as shown below.

Estimating glitch generation in the controller

Glitch generation in the control logic can be thought to be a product of the interaction of certain logic (values assumed by the gate's inputs) and temporal (timing relationship between events at the gate's inputs) conditions, as illustrated by the following example.

Example 3.5 *Consider the RTL circuit shown in Figure 3.5 once again. Let us focus on control signal $contr[2]$, which is highly glitchy according to the activity profiles of Figure 3.14. The portion of the decode logic that implements this control signal is shown in Figure 3.15(a). The total and zero-delay switching activities are indicated on each line. A $1 \rightarrow 0$ or $0 \rightarrow 1$ transition is counted as a half-transition. Though the inputs are largely glitch-free, significant glitches are generated at AND gates $G1$ and $G2$. After careful analysis, the generation of glitches was attributed to two conditions:*

- **C1:** *A rising transition on signal $x1$ is frequently accompanied by a falling transition on $c11$. Thus, the rising transition on $x1$ and the falling transition on $c11$ are highly correlated.*

- **C2:** *Transitions on signal $x1$ arrive earlier than transitions on signal $c11$.*

Condition C1 arises due to the functionality of the design: most of the times when state $s1$ is entered (rising transition on $x1$), the comparisons evaluated by the comparators feeding $c9$ and $c10$ evaluate to 0, changing from 1 in the previous state. On the other hand, condition C2 is a result of the temporal characteristics of the design. These conditions, captured graphically in Figure 3.15(b), lead to the generation of glitches at gate $G1$, that propagate to control signal $contr[2]$. A similar explanation holds for the output of gate $G2$ being glitchy.

In general, the *logic conditions* necessary for glitch generation at a gate during an interval of time are as follows.

- There should be at least one rising and at least one falling transition at the gate's inputs.
- No input should assume a steady controlling logic value (*e.g.* 0 for AND, 1 for OR, *etc.*) throughout the interval under consideration.

Assuming an inertial delay model, the temporal condition for glitch generation in an AND gate is as follows.

- The earliest falling transition arrives after the latest rising transition by an interval that is greater than the gate's inertial delay.

Similar conditions can be derived for glitch generation in other types of gates.

Given a control expression in a sum-of-products form, as shown in Equation (3.21), the glitching activity generated due to each product term is estimated separately, as well as glitches generated due to the disjunctive composition of the product terms. A useful property of control expressions that is utilized in making the above approximation is the fact that the product terms are typically disjoint since they are derived to represent distinct combinations of the controller state and condition values.

ARCHITECTURE-LEVEL POWER ESTIMATION   73

The logic conditions for glitch generation for each product term (conjunctive or AND expression) and the disjunctive (OR) expression combining the product terms can be easily monitored during zero-delay RTL simulation. A distinct *glitch counter* is maintained for each product term, and also for the OR expression combining the product terms. In each simulation cycle, the previous and current values at the variables involved in an AND or OR expression are checked to see whether the logic conditions for glitch generation are satisfied. If they are, the corresponding glitch counter is incremented to indicate the possibility of glitch generation in the current simulation cycle.

As mentioned earlier in this section, checking whether the temporal conditions for glitch generation are satisfied in an accurate manner requires the final implementation of the control logic, which is typically not available when performing high-level design optimizations. High-level timing analysis techniques [66, 91] can only provide a rough estimate, or a bound on the actual time at which a signal makes its last transition in a clock cycle. For accurately predicting the generation of glitches at a gate, it is necessary to know the exact times at which each of the gate's inputs makes a transition, if any, in each clock cycle. One possible approach to tackle the lack of accurate delay information is to make a pessimistic assumption, *i.e.* assume that glitches are generated at a gate whenever the logic conditions for glitch generation are satisfied. However, in practice, this pessimistic assumption often leads to substantial over-estimates of glitches at control signals, as shown in the following example.

Example 3.6 *Consider the control signal $contr[2]$ in the* GCD *RTL circuit of Figure 3.5. The control expression for $contr[2]$ is $x0 + x1.c11 + x3.c10$. Suppose the aim is to estimate the glitching activity at control signal $contr[2]$, given the traces for each of the decoded state and comparator output signals that were captured during zero-delay RTL simulation. In this case, signals $c10$ and $c11$ were found to be glitch-free, simplifying the problem to that of estimating glitch generation at $contr[2]$.*

74 HIGH-LEVEL POWER ANALYSIS AND OPTIMIZATION

For the time being, let us make pessimistic assumptions to tackle the lack of availability of complete temporal information, i.e. we conclude that glitches are generated whenever the logic conditions for glitch generation are satisfied. Clearly, the first product term ($x0$) cannot generate any glitches. From the simulation traces, the following statistics were computed for logic conditions for glitch generation at the second and third product terms.

$$Case\ 1: Count(x1\downarrow, c11\uparrow) = 15,\ Case\ 2: Count(x1\uparrow, c11\downarrow) = 20,$$
$$Case\ 3: Count(x3\downarrow, c10\uparrow) = 35,\ Case\ 4: Count(x3\uparrow, c10\downarrow) = 30$$

In the above equations, the symbols $\uparrow$ and $\downarrow$ denote rising and falling transitions, respectively. The expression $Count(x1\downarrow, c11\uparrow)$ represents the number of instances (consecutive pairs of cycles) in the simulation trace where $x1$ makes a falling transition, while $c11$ simultaneously makes a rising transition. From the above numbers, one could conclude that the glitching activity generated due to the second and third product terms is 35 and 65, respectively. As explained previously, the glitches generated due to each product term propagates to the output un-mitigated, since the decoded state variables are mutually exclusive. From the given traces, it was observed that the logic conditions for glitch generation at an OR *gate were never satisfied by the outputs of the product terms. Hence, the glitching activity at control signal $contr[2]$ was estimated to be 100 transitions over the entire simulation period. A comparison with the glitching activity observed for the same input traces by a gate-level power simulator after obtaining the implementation shows that the glitching activity at $contr[2]$ was over-estimated by the pessimistic approach by as much as 92.3%.*

Although exact arrival time information at various signals is not available, it is often possible to derive *partial information* about delays from RTL descriptions or during high-level synthesis. For example, the outputs of comparators can often be assumed to arrive later than the decoded present state signals, even when knowledge of their exact arrival times is not available. In essence, partial delay information can be thought of as a set of relationships between the arrival times at two or more inputs of the controller. Partial delay information can be combined with the logic conditions for glitch generation to refine the glitch estimates at control signals as shown next. Inputs to the control logic are divided into three groups - *early* arriving signals, *late* arriving signals, and signals whose arrival time information is assumed to be *unknown*. Each controller input signal that is marked as late-arriving is assumed to arrive significantly later than any input signal that is marked as early-arriving. No assumption is made involving the arrival time of a signal marked unknown. Similarly, no assumption is made about the relationship between the arrival times of two signals that are both marked either early or late. When the temporal conditions for glitch generation at a gate involve signals whose arrival time relationship is unknown, the pessimistic approach of only checking logic conditions is used.

Example 3.7 *Let us revisit control signal $contr[2]$ in the* GCD *circuit that was used for the discussions in Example 3.6. Suppose the comparator output signals, $c10$ and $c11$, arrive after the decoded state variables, $x0$, $x1$ and $x3$. Consider Case 1 ($x1 \downarrow, c11 \uparrow$) in the equations presented before. Since the rising transition arrives later than the falling transition in this case, the temporal conditions for glitch generation are not satisfied for this case. Similarly, it can be seen that Case 3 does not satisfy the temporal conditions for glitch generation in the third product term. However, they are satisfied for Case 2 and Case 4. The revised glitching activity estimate for $contr[2]$ is, therefore, 50, which represents an error of only 4% with respect to the number reported by a gate-level power simulator.*

Glitch propagation through the control logic

This subsection explains the procedures for estimating glitch propagation through the control logic, *i.e.* from comparator outputs to control signals. Consider again the generic control expression given in Equation (3.21). Consider a particular comparator output, $c1$, that has been predicted to be glitchy based on the data path glitching activity models.

In order for glitches at $c1$ to propagate to the control signal, at least one of the product terms it is involved in must have non-controlling side inputs (*i.e.* 1), and the result of all other product terms should evaluate to 0. Hence, the following equation can be utilized to estimate the propagation of glitches to the control signal.

$$Gl(c1) * P\left(\frac{\partial contr}{\partial c1}\right) \qquad (3.22)$$

In the above equation, $Gl(c1)$ represents the glitching activity at $c1$, and $\frac{\partial contr}{\partial c1}$ is the Boolean difference [5] of the control signal with respect to $c1$. The term $P\left(\frac{\partial contr}{\partial c1}\right)$ in the above equation can be thought of as the probability that the control signal will be "sensitized" to glitches at $c1$. This probability can be computed easily during RTL simulation. Alternatively, it is possible to use known accurate probability propagation techniques that are applicable to two-level expressions.

Example 3.8 *Consider control signal $contr[5]$ of the GCD RTL circuit. The control expression for $contr[5]$ is $x1.\overline{c11}.c15 + x2.c15 + x3.\overline{c10}.c15$. Let us focus on the propagation of glitches from $c15$ to $contr[5]$ through the product term $x2.c15$. Using Equation (3.22), and the observation that the product terms are disjoint, the contribution of the product term of interest can be written as $Gl(c15) * P(x2 = 1)$. From the RTL simulation traces, the value of $P(x2 = 1)$ was computed to be 0.647. Combining this number with the estimated glitching activity at $c15$ results in an estimate of 27.3 (cumulative activity for the entire simulation run) for the glitching activity due to the term $x2.c15$. However, the glitching activity reported by the gate-level power simulator for the entire simulation run was only 1.0. Upon further investigation into this discrepancy, it was observed that the conditions for glitch generation at signal $c15$ were negatively correlated with the conditions for glitch propagation through the chosen product term. In other words, in consecutive pairs of cycles in which the controller made a state transition into state $s2$, the data inputs of the comparator were such that the glitches at the comparator's output*

---

[5] The Boolean difference of a function $f(x_1, \ldots, x_n)$ with respect to a variable x_i is $f_{x_i} \oplus f_{\bar{x}_i}$, where $f_{x_i} = f(x1, \ldots, x_{i-1}, 1, x_{i+1}, \ldots, x_n)$ and $f_{\bar{x}_i} = f(x1, \ldots, x_{i-1}, 0, x_{i+1}, \ldots, x_n)$.

were minimal, leading to almost no propagation of glitches from $c15$ through the product term $x2.c15$.

The above problem is resolved by predicting the glitches at $c15$ separately *for each state. For example, in order to predict the glitch propagation for the product term* $x2.c15$, *the glitching activity at* $c15$ *is estimated for only those consecutive pairs of cycles where the final controller state is* $s2$. *Since the glitches at* $c15$ *are being decomposed into separate estimates for each state, this technique is referred to as* glitching activity decomposition. *Glitching activity decomposition exposes any correlations between the conditions required for glitch generation and those required for glitch propagation, leading to an improvement in the accuracy of the glitch estimates. Note that in order to compute a separate figure for glitching activity at* $c15$ *in state* $s2$, *it is necessary to compute separate figures in state* $s2$ *for the zero-delay statistics and glitching activity at the inputs of the comparator that generates* $c15$. *In practice, decomposition does not impose a significant computational bottleneck, since (i) the decomposition of statistics by state is limited to only the transitive fanins of those comparators that were found to generate glitches, and (ii) separate statistics are computed only for those states that are related to a glitchy comparator output through a product term. In the current example, it was found that when the controller made a state transition to state* $s2$, *the temporal correlation at the inputs of the comparator was very high, leading to minimal generation of glitches in* $s2$. *The predicted glitching activity at the output of the* AND *gate implementing* $x2.c15$ *now becomes* 3.0, *which is much closer to the gate-level estimate.*

In order to get a feel for the accuracy of the switching activity estimation techniques for control signals, results are provided comparing the estimates to the switching activity measured after a complete gate-level implementation using a gate-level power simulator [81], for all distinct control signals in the GCD RTL circuit (except $contr[1]$, which is not glitchy). The scatter plot shown in Figure 3.16 shows the results of this experiment. The x-coordinate represents the total switching activity reported by the gate-level simulator for the control signal, while the

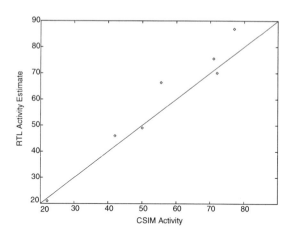

Figure 3.16. Scatter plot of switching activity at control signals: RTL estimate *vs.* gate-level estimate

y-coordinate represents the switching activity estimated using the RTL activity estimation procedure. As a reference, the plot also shows a solid line for the equation $y = x$, *i.e.* points in the scatter plot close to this line indicate a high accuracy in the RTL estimates. The figure indicates that the presented control signal switching activity estimation techniques produce estimates that are quite close to the activity numbers obtained at the gate-level after a time-consuming implementation of the complete GCD controller and data path.

3.4 CONCLUSIONS

This chapter presented several techniques for estimating the power consumption at the register-transfer or architecture level. In order to obtain a high degree of estimation accuracy and efficiency, it is necessary to use a variety of techniques, including different modeling techniques for different parts of a design (such as arithmetic macroblocks, control logic, memory, clock, and I/O). Sometimes it may be desirable to use specific techniques depending on the applications. However, the presence of this diversity of techniques makes it challenging to develop tools that are applicable to a large class of designs by employing the appropriate technique depending on the characteristics of the design and its environment, as well as the part of the design being analyzed. An important direction of future work should be to combine the various techniques in a seamless manner.

4 POWER MANAGEMENT

Power consumption in CMOS circuits is dominated by the dynamic component that is incurred whenever signals in the circuit undergo logic transitions. In practice, a large portion of the signal transitions that occur in a circuit are unnecessary, *i.e.* they have no effect on the value at the circuit output. Recognizing this fact, several techniques have been proposed to reduce power consumption by eliminating unnecessary transitions at various signals in the circuit. The term power management is used to collectively refer to such techniques that exploit the fact that not all parts of a circuit are needed to function in each clock cycle. Such techniques identify conditions under which various parts of the circuit are idle, and shut them down to reduce power consumption. Power management is a concept that can be applied at each level of the design hierarchy. This chapter covers several power management techniques that can be applied during high-level design, including gated and multiple clocks, pre-computation, scheduling for power management, operand isolation, constrained register sharing, and controller-based power management.

82   HIGH-LEVEL POWER ANALYSIS AND OPTIMIZATION

4.1 CLOCK-BASED POWER MANAGEMENT: GATED AND MULTIPLE CLOCKS

The technique of gating clocks is probably the most well-known and most commonly used form of power management. The idea is to suppress or disable transitions from propagating to parts of the clock network under specific conditions that are determined by the clock gating circuitry. Clock gating can result in savings due to reduced capacitive switching in the clock network, that may include clock buffers, the interconnect of the clock network, and the latches/registers that are fed by the clock signal. In addition, gating the clock may prevent storage elements from loading unnecessary new values, thus leading to savings in power consumption in the logic fed by the registers.

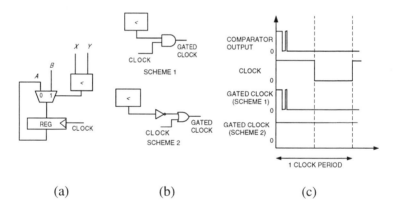

Figure 4.1.   Gating clock signals to save power

The technique of gating clocks to registers is illustrated in Figure 4.1. In the circuit in Figure 4.1(a), the register re-loads its previous value when the less-than comparator's output is 0. Hence, whenever the comparator output evaluates to 0, the transition at the clock input to the register can be suppressed. For this example, assume that the design is based on rising edge-triggerred flip-flops. Figure 4.1(b) shows two candidate schemes, *Scheme* 1 and *Scheme* 2, to gate the clock input to the register. The rationale behind *Scheme* 1 is that the register's clock input would be forced to a 0 whenever the output of the comparator evaluates to a 0, thus suppressing unnecessary transitions on the clock. In *Scheme* 2, the register clock input is forced to a 1 whenever the comparator's output evaluates to 0. Thus, an initial analysis may suggest that both the schemes are equivalent. However, it is also necessary that certain timing constraints be satisfied by the signal used to gate the clock. Consider the sample waveforms shown in Figure 4.1(c) for both the schemes. For *Scheme* 1 to work, we require that the comparator's output evaluate to 0 before the clock edge rises, which is not possible to satisfy. Hence, *Scheme* 1 does not work when timing considerations are taken into account. On the other hand, *Scheme* 2 will work as long as the gating condition stabilizes before the clock signal makes a high-to-low transition.

Often, existing signals in the circuit can be used for gating a part of the clock network. Such signals may be present in the same clock cycle as they are used, or may be available in previous clock cycles in which case latches need to be inserted to save them till they are used. For example, signals available in the instruction decode stage(s) of a microprocessor pipeline may be used for clock gating in later pipeline stages. Otherwise, circuitry needs to be added to the design in order to generate the clock gating signals. While clock gating is a very useful technique for saving power, there are several pitfalls and overheads incurred that need to be considered, as follows.

- Introducing gates into the clock tree can lead to an increase in clock delay and clock skew.

- Failure to ensure that the gating logic does not introduce glitches on the clock signal can lead to circuit malfunction due to spurious loading of registers.

- Circuits with gated clocks present significant additional complexity to synthesis and analysis tools.

84   HIGH-LEVEL POWER ANALYSIS AND OPTIMIZATION

4.1.1 Automatic synthesis of gated-clock circuits

Given a finite-state machine (FSM) description of a sequential circuit, it is possible to disable the entire clock network by identifying conditions when the next state and primary output values do not change. Gating the clock tree at its root eliminates the clock skew problems that can be introduced when only parts of the clock tree are gated. The circuitry added to identify all such idle conditions may incur excessive power, delay, and area overheads. Hence, it is important to develop techniques to detect a subset of the idle conditions at significantly reduced overheads. Automated gated-clock FSM synthesis techniques for Moore FSMs were presented in [92], and extended to a more general class of FSMs in [93]. This procedure assumes FSMs with registered primary inputs, as shown in Figure 4.2, where the state and input registers are combined into a single register. Clock gating is achieved by synthesizing an *activation function*, F_a, which evaluates to logic 1 when the clock needs to be stopped. The latch L, which is transparent when the clock signal is low, ensures that glitches are not propagated onto the clock signal, and allows simpler delay constraints to be imposed on the output of the circuit realizing F_a. The AND gate suppresses the clock transition when the latch output is 1.

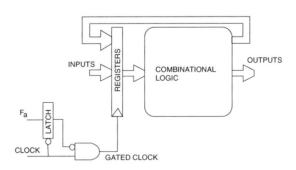

Figure 4.2.   Gated-clock FSM architecture

The procedure for synthesizing F_a is based on the following observation. Consider a Moore FSM, $(PI, PO, S, s0, \delta, \lambda)$, where PI is the set of inputs, PO

is the set of outputs, S is the set of states, $s0$ is the initial state, δ is the next-state function, and λ is the output function. For Moore FSMs, the output is a function only of the current state, and not of the input variables. A self-loop in the state transition graph (STG) corresponds to an idle condition, *i.e.* a condition when the clock to the FSM register can be suppressed. Let us define for each state si a self-loop function, $Self_{si} : PI \rightarrow \{0,1\}$ such that $Self_{si}(pi) = 1$ if and only if $\delta(x, si) = si$, $pi \in PI$. $Self_{si}$ captures the set of input conditions under which the self-loop of state si is traversed. The activation function F_a can now be written as

$$F_a = \sum_{i=0,\ldots,|S|-1} Self_{si}.xi \qquad (4.1)$$

where xi is the decoded state variable corresponding to state si, *i.e.* xi is 1 if and only if the FSM is in state si. Mealy FSMs can be handled by transforming them into *locally Moore* FSMs by splitting states with self-loops that have different output labels on their incoming edges into multiple states which have only one output label on incoming edges, and applying the same procedure.

It is possible that the implementation of F_a may itself consume significant power and outweigh the savings obtained by disabling the clock signal. Hence, it may sometimes be desirable to identify a proper subset of the FSM's idle conditions that are most probable so that the implementation complexity of F_a is reduced. This is approximated in [93] by solving a *constrained-probability minimum literal-count covering* problem where the aim is to obtain a sub-function of F_a whose probability of evaluating to a 1 is greater than a pre-specified threshold, and the number of literals in the implementation of the sub-function is minimal.

The above scheme of disabling the entire clock tree is well-suited to FSM designs where the complete FSM is idle for significant amounts of time, *e.g.* reactive systems that wait for long periods of time for an input event to occur before they produce a response. In more general designs, where the complete clock network cannot be disabled, it may still be possible to shut down smaller sub-networks that feed idle parts of the circuit. However, the above procedure is inapplicable to

86   HIGH-LEVEL POWER ANALYSIS AND OPTIMIZATION

circuits or sub-circuits that cannot be efficiently represented as an FSM, *e.g.* data path circuits.

4.1.2   Clock gating techniques for data path registers

For data path registers in RTL circuits, gating conditions for registers and sets of registers may be obtained through a symbolic analysis of the combinational circuit that feeds it [94]. The aim is to determine the conditions under which the register retains or re-loads its value. It is common in manual designs as well as designs produced by high-level synthesis to have a register fed by a multiplexer network, where the register's output is fed back as one of the data inputs to the multiplexer network. The conditions under which such a self-loop is logically activated represent the conditions under which the register retains its previous data value. These conditions are identified by traversing the path through the multiplexer network that forms the register self-loop. The condition for this path to be activated is computed in terms of the select signals connected to the individual multiplexers along the path. The condition that the path is activated can be written as the *conjunction* of the conditions that each multiplexer along the path selects the on-path input. This technique could be extended to also handle self-loops that pass through other circuit blocks, such as functional units, by symbolically analyzing the conditions under which the self-loop propagates the register's old value.

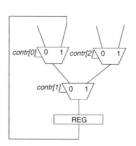

Figure 4.3.   Deriving clock gating conditions for data path registers

Example 4.1 *Consider the register and multiplexer tree feeding it shown in Figure 4.3. Assuming that we are using $Scheme$ 2 shown in Figure 4.1, the gating condition for the clock input to the register is $contr[0].\overline{contr[1]}$.*

Next, consider the general case, where a register in an RTL circuit has a self-loop passing through n 2-to-1 multiplexers in a multiplexer network. Let $Sel_1, Sel_2, \ldots, Sel_n$ represent the conditions under which the multiplexers in the path that forms the self-loop select the on-path inputs. Sel_i is either equal to the control signal that feeds the select input of the corresponding multiplexer, or its complement, depending on whether the on-path input is the 1-input or 0-input of the multiplexer. The gating condition for the register clock can then be written as $Sel_1.Sel_2\ldots Sel_n$.

Since the logic to compute the select signals of the various multiplexers in the multiplexer network is already present in the circuit, the only logic that needs to be added is that required to invert the control signals, where necessary, and compute the conjunction of the multiplexer select signals or their complements. The above procedure to derive gating conditions does not guarantee that the required timing constraint (the gating condition should stabilize before the clock signal makes a high-to-low transition) is met. In order to avoid slower clocking of the design or changing the duty cycle of the clock (by delaying the falling edge of the clock), the above procedure can be augmented as follows. If the timing constraint is violated by the gating condition, computed as explained in the previous paragraph, a *reduced* gating condition is derived using a subset of the terms used in the original gating condition. The expression for the gating condition is first collapsed to a two-level sum-of-products form [47], and a high-level delay estimator, such as FEST [66], is used to determine the arrival times at the signals representing each product term. A subset of the product terms is identified such that the largest arrival time among the product terms plus the delay of the logic required to compute the OR of the selected terms is less than the duty cycle. This could be performed using procedures similar to those used for the constrained-probability minimum literal-count covering problem [93], with the difference that the aim is to eliminate

88 HIGH-LEVEL POWER ANALYSIS AND OPTIMIZATION

terms that violate the timing constraint as opposed to minimizing the power cost required to implement the extra logic.

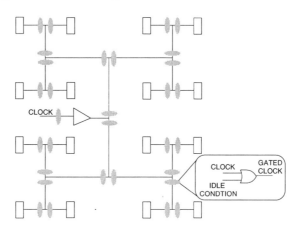

Figure 4.4. Clock gating at multiple levels in the clock tree

While the above procedure derives a separate gating condition for each data path register, it is possible to combine the gating conditions for a group of registers into a single gating condition that can be used to gate the clock input to all the registers in the group. The benefit of such merging is that it is possible to suppress unnecessary transitions in larger parts of the clock distribution network. In general, it is possible to gate the clock signal at various points in the clock distribution network. Consider the clock tree shown in Figure 4.4, that has disabling gates at every level in the clock tree. The choice of gating the clock at various points in the clock tree leads to a trade-off between the physical capacitance that is prevented from switching, and the number of unnecessary transitions avoided. Disabling the clock signal higher up in the clock tree implies that a larger capacitance is prevented from switching. However, the clock transition at a point in the clock tree can be suppressed only if all the registers fed by the sub-tree rooted at that point re-load or retain their old value. Hence, the gating condition is satisfied fewer times, resulting in a reduction in the number of transitions saved in the clock tree. In other words, the number of transitions suppressed at the clock inputs to some of the registers, when we use a merged gating condition, may be less than the number of transitions that could have been suppressed by using individual gating conditions. It may sometimes be beneficial to gate the clock tree at multiple levels, *i.e.* have multiple gating points along a path.

4.1.3   Clock tree construction to facilitate clock gating

The manner in which the clock tree is constructed determines the effectiveness of gating the clock tree at higher levels. For example, consider two candidate clock trees constructed to feed four registers, $R1, \ldots, R4$, as shown in Figure 4.5. Each register is marked with the conditions under which its clock signal can be disabled. $x1, \ldots, x4$ represent decoded controller state variables and are hence mutually exclusive, *i.e.* no two of those signals can assume a value of 1 simultaneously. Consider the clock tree shown in Figure 4.5(a). $R1$ and $R2$ are grouped together at the lowest level of the tree. Since the conditions under which $R1$ and $R2$ can be gated are mutually exclusive, their conjunction can never be true. Hence, it is not possible to gate the clock signal at the point marked A in Figure 4.5(a). A similar argument applies to $R3$ and $R4$. Now consider the alternative clock tree shown

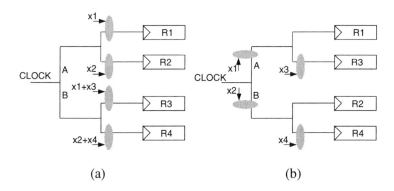

Figure 4.5.   Effect of clock tree structure on clock gating possibilities

in Figure 4.5(b). The gating condition for the sub-tree rooted at point A can be computed as

$$GC_A = x1.(x1 + x3) = x1$$

Similarly, the gating condition for point B can be computed to be $x2$. Thus, the clock tree shown in Figure 4.5(b) is more suited to clock gating, since it *groups registers with similar or overlapping idle conditions closer together* in the clock tree. Therefore, the scope that a clock tree allows for power management is an additional parameter that must be taken into account during clock tree construction. Techniques for addressing this problem were presented in [95].

4.1.4 Power management using multiple non-overlapping clocks

The use of gated clocks results in the clock signals, that feed various sub-circuits, being suppressed when the registers in the sub-circuits do not need to load a new value. The cycles during which the clock transitions are suppressed need not follow any regular pattern in general, since the suppression of clock signal transitions is data-dependent. Some types of designs, however, contain sub-circuits whose idle clock cycles follow a simple, regular pattern. For example, a component may be active and idle in alternating clock cycles. If the cycles in which a sub-circuit is idle follows a regular pattern, the clock generation circuitry need not be data-dependent. For example, consider a circuit that consists of two sub-circuits A and B. A and B are each active for one clock cycle and idle for the next clock cycle in an alternating fashion such that exactly one of them is active in a given clock cycle. Suppressing the unnecessary transitions on the clock inputs to A and B results in the waveforms which can be generated by using multiple non-overlapping clocks. Thus, the use of multiple slower clocks in a design instead of a single faster clock is in itself an implicit method of implementing power management. Power consumption may be reduced in the clock circuitry since each of the clock trees has a smaller activity factor than a single clock, and the aggregate physical capacitance of the various clock trees feeding smaller portions of the circuit is not much higher than the capacitance of a single clock tree that feeds the entire circuit.

Example 4.2 *Consider a circuit where all the registers are fed by a single clock network that has physical capacitance C and switches at frequency f. Suppose the design is partitioned into two parts fed by clock signals that have frequencies f/2, and physical capacitances C_1 and C_2 respectively. Power savings in the clock network results from the use of two separate clocks if the following condition holds.*

$$C_1 \cdot \frac{f}{2} + C_2 \cdot \frac{f}{2} < C * f \tag{4.2}$$

In other words, the use of multiple clocks leads to power savings in the clock network as long as $C_1 + C_2 < 2 * C$. This condition is likely to hold if the design is partitioned carefully so that the registers that need to be fed by each clock are clustered close together in the layout.

While the above example illustrates how the use of multiple clock signals may save power in the clock network, power savings may also result in the other parts of the circuit, due to a reduction in unnecessary switching activity.

Techniques for high-level synthesis of multiple non-overlapping clock based designs were presented in [96]. Consider the scheduled DFG shown in Figure 4.6(a). The clock cycles of the schedule, $s1, \ldots, s5$, have been assigned to two non-overlapping clock domains, $CLOCK1$ and $CLOCK2$, in an alternating fashion. Figure 4.6(b) shows a single clock RTL circuit, that implements the given DFG using minimal resources, that is derived without considering the clock partitions shown in Figure 4.6(a). Figure 4.6(c) shows an RTL circuit that has been implemented with the following additional resource sharing restrictions:

- An operation scheduled in a clock cycle assigned to $CLOCK1$ cannot share a functional unit with an operation scheduled in a clock cycle assigned to $CLOCK2$.

- A variable that is generated in a clock cycle assigned to $CLOCK1$ cannot share a functional unit with a variable that is generated in a clock cycle assigned to $CLOCK2$.

92   HIGH-LEVEL POWER ANALYSIS AND OPTIMIZATION

The above restrictions ensure that each register in the design can be clocked by either $CLOCK1$ or $CLOCK2$. In addition, the data path can be partitioned into two domains such that there is switching activity in the partitions or clock domains only during their respective active clock cycles.

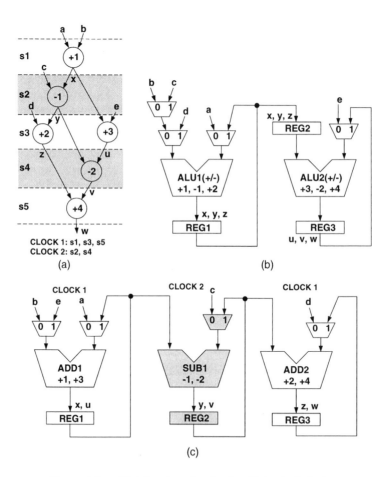

Figure 4.6.   High-level synthesis of multiple clock designs

4.2 PRE-COMPUTATION

Pre-computation [97] is a power management technique that involves selectively computing one or more cycles in advance (or pre-computing), using much simpler circuitry than the original circuit itself, the output values of the circuit, and using the pre-computed values to reduce internal switching activity in the succeeding clock cycle. Consider the circuit shown in Figure 4.7(a), that consists of a combinational logic block A sandwiched between registers $R1$ and $R2$ that have load enable signals $LE1$ and $LE2$. Let us assume that block A has a single output. The circuit with pre-computation logic is shown in Figure 4.7(b). The *predictor* functions, g_1 and g_2, satisfy the following conditions:

$$g_1 = 1 \Rightarrow f = 1$$
$$g_2 = 1 \Rightarrow f = 0$$

During clock cycle n if either g_1 or g_2 evaluates to a 1, register $R1$ is disabled from loading. As a result, the values at the inputs to block A do not change, resulting in savings in power consumption. Note that the extra logic added to the circuit

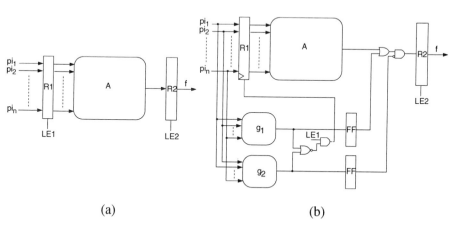

Figure 4.7. (a) Original circuit, and (b) circuit after applying pre-computation

incurs area, delay, and power overheads. The delay overhead can be avoided by applying this technique to non-critical sub-circuits. The area and power overheads can be minimized by reducing the complexity of functions g_1 and g_2. The goal is to include as many input conditions as possible in g_1 and g_2, *i.e.* maximize the signal probabilities of g_1 and g_2, while ensuring that g_1 and g_2 are significantly less complex than the original function itself. The pre-computation architecture shown in 4.7(b) is called the *complete input disabling* architecture, since all the inputs to the combinational logic block A are disabled from changing when either g_1 or g_2 evaluates to a 1. The same concept can also be applied to disable a subset of the inputs to a block, as shown in the following example.

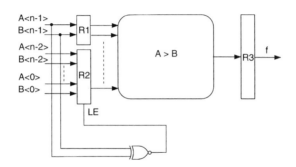

Figure 4.8. Input subset disabling through pre-computation

Example 4.3 *Consider an n-bit comparator shown in Figure 4.8 that computes the function $A\langle n-1,\ldots,0\rangle > B\langle n-1,\ldots,0\rangle$. The most significant bits of A and B, when they are not equal, are sufficient to determine the result of the comparison. Thus, we can have the predictor functions depend only on $A\langle n-1\rangle$ and $B\langle n-1\rangle$. The circuit with the pre-computation logic is shown in the figure. The expressions for the predictor functions are:*

$$\begin{aligned} g_1 &= A\langle n-1\rangle.\overline{B\langle n-1\rangle} \\ g_2 &= \overline{A\langle n-1\rangle}.B\langle n-1\rangle \end{aligned}$$

When either of the predictor functions evaluates to 1, all inputs to the comparator except the most significant bits can be disabled. The reduced switching activity at the inputs of the comparator leads to savings in power consumption. This scheme can be extended by also using $A\langle n-2\rangle$ and $B\langle n-2\rangle$ for computing the predictor functions, and disabling the remaining bits of A and B, and so on.

4.3 SCHEDULING TO ENABLE POWER MANAGEMENT

Logic-level shut-off techniques, such as pre-computation, are based on disabling the input latches of a module when the output of the module is not used, on a per-clock-cycle basis. Such techniques are limited by the inherent logic structure of the circuit. Looking at higher-level representations of a design, such as a behavioral description or CDFG, makes it possible to perform power management trade-offs that are more global in nature. For example, the result of conditional operations in CDFGs determine which parts of the CDFG are executed and which are not. A common performance optimization technique is to execute the operations for all the outcomes of a conditional in parallel with the computation of the condition itself, and choose the appropriate results based on how the condition evaluates [44]. This strategy is, however, not well-suited to power optimization, since unnecessary operations may be executed. For power management, it is desirable to enforce the control dependencies between operations in the CDFG and the conditional operations that they depend on during scheduling. This may be done in a selective manner, in order to meet the performance constraints while minimizing the number of unnecessary operations executed. Scheduling techniques to explore such trade-offs were presented in [98].

Example 4.4 *Consider the computation of the expression $|a - b|$. The CDFG for this computation is shown in Figure 4.9(a). Suppose that each of the operations ($-$, $>$) takes one clock cycle, and that the select (Sel) operation can be chained with any of the other operations. We want to obtain an implementation that performs*

96   HIGH-LEVEL POWER ANALYSIS AND OPTIMIZATION

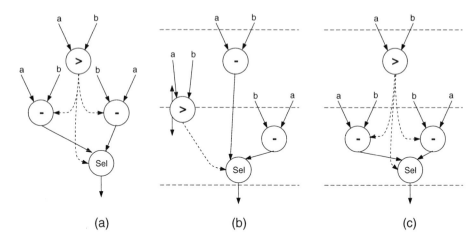

Figure 4.9.   Scheduling to enable power management

the computation within two clock cycles. Two possible schedules that meet the performance constraint are shown in Figure 4.9(b) and Figure 4.9(c). The schedule of Figure 4.9(b) uses the so-called D-select *conditional representation [44], where the control dependencies from the ($>$) operation to the ($-$) operations are ignored. The computations of $a-b$ and $b-a$ are performed independently of the comparison, and the result of the comparison is used by the Sel operation, which may be implemented using a multiplexer, to select the appropriate result. Note that there is a flexibility in scheduling the ($>$) operation in either of the two control steps. In general, this flexibility can be utilized in order to either implement the design using fewer resources or to improve the number of clock cycles in the schedule. From the point of view of power consumption, however, the schedule of Figure 4.9(b) is wasteful, since both $a - b$ and $b - a$ are computed although only one of them is eventually used. In the RTL implementation, if the two ($-$) operations are assigned to different subtracters, the inputs to both the subtracters will change, causing unnecessary power dissipation in both. On the other hand, if both ($-$) operations are assigned to the same subtracter, the inputs to the subtracter changes* in both the clock cycles, causing power dissipation in the subtracter in both clock cycles. This can be avoided by scheduling operations that determine the control flow as early as possible, and using their results to activate other resources only when necessary. For example, in the schedule of Figure 4.9(c), the (>) operation is assigned to the first control step, and depending on its result, only $a - b$ or $b - a$ is performed. In the RTL implementation, if the two (−) operations are assigned to two different subtracters, only the inputs to one of the subtracters is allowed to change, depending on the result of the conditional. If the two (−) operations are assigned to the same subtracter, the subtracter's inputs do not change in the first clock cycle, and are set appropriately in the second clock cycle depending on how the conditional evaluates.

Techniques to decide which conditionals to apply power management to in order to obtain maximal power savings, while meeting the specified performance, were presented in [98].

4.4 OPERAND ISOLATION

The techniques presented in Sections 4.1 and 4.2 are only applicable to blocks of combinational logic that are fed by registers, *i.e.* they are not applicable to circuit blocks embedded within combinational logic. *Operand isolation* [99, 100, 101, 102] is a technique that can be used to save power consumption in idle circuit blocks by disabling transitions at their inputs. Operand isolation is illustrated in Figure 4.10. Transparent latches are inserted at all the inputs of an embedded logic block, and control circuitry is added to detect the idle conditions for the block. When the block is not required to perform any useful operation, the transparent latches at its inputs are disabled, and retain the previous cycle's values, avoiding unnecessary power dissipation in the idle block.

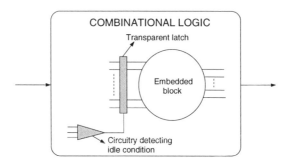

Figure 4.10. Operand isolation

4.4.1 Guarded evaluation

An automated technique, called *guarded evaluation*, that detects idle sub-circuits on a per-clock-cycle basis, and inserts transparent latches to perform operand isolation was presented in [100]. Let o be a signal in a combinational circuit. Let F represent the logic that is needed to compute o and no other signal. I is the set of inputs to F. Let ODC_o refer to the set of primary input assignments to the entire circuit such that the value of o has no influence on the values at the primary outputs. In other words, ODC_o is the observability don't care set [47] of signal o. Let LE be any other signal in the circuit. Let $t_e(I)_{LE=1}$ represent the earliest

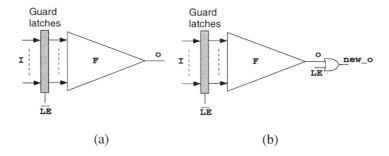

Figure 4.11. Guarded evaluation

time at which any of the inputs in I can change its value when $LE = 1$. Let $t_l(LE)_{LE=1}$ represent the latest time at which LE can stabilize to logic value 1. Signal LE can be used to *guard* the logic block F, as shown in Figure 4.11(a), if the following conditions are satisfied:

- $LE \Rightarrow ODC_o$, i.e. $\bar{LE} + ODC_o \equiv 1$

- $t_l(LE)_{LE=1} < t_e(I)_{LE=1}$

The first condition ensures that the circuit does not need to compute o when $LE = 1$. The second condition ensures that the transparent latches are disabled in time, *i.e.* early enough to cut off transitions on any of the inputs in I. The architecture shown in Figure 4.11(a) is referred to as *pure guarded evaluation*, since no extra logic is added to the circuit except the transparent latches and that required to invert signal LE, if necessary.

The above idea can be extended as shown in Figure 4.11(b). Suppose that signal LE satisfies the relaxed logic condition $LE \Rightarrow (o + ODC_o)$, i.e. $\bar{LE} + o + ODC_o \equiv 1$. The temporal condition ($t_l(LE)_{LE=1} < t_e(I)_{LE=1}$) is the same as above. Under the relaxed logic condition, when LE evaluates to 1, either of the following two cases may hold: (i) o is not needed to compute the value at the primary outputs, or (ii) o is needed to compute the primary outputs, however, o should evaluate to a 1. The circuit of Figure 4.11(a) may operate incorrectly under the second case, since o will assume the previous cycle's value which may or may not be 1. In order to ensure correct operation, an OR gate is added that forces the output of F to 1 when $LE = 1$, as shown in Figure 4.11(b). This technique is referred to as *extended guarded evaluation*. The case when $LE \Rightarrow (\bar{o} + ODC_o)$ can be similarly exploited by using an AND gate instead of an OR gate. The logic condition can be verified using binary decision diagram [103, 104] based techniques, or automatic test pattern generation [105] based techniques.

4.4.2 Operand isolation in the context of high-level synthesis

The idea of operand isolation can be easily applied to RTL circuits generated by high-level synthesis tools [102]. Since high-level synthesis involves assigning operations in a behavioral description to clock cycles or controller states (scheduling), and mapping operations and variables to resources such as functional units and registers (resource sharing), the conditions under which a resource (*e.g.* a functional unit) is not used are readily available from the scheduling and resource sharing information. For example, consider the scheduled DFG shown in Figure 4.12, and the corresponding RTL implementation shown in Figure 4.13. The RTL circuit consists of functional units, registers, multiplexers, and a controller FSM. From the schedule, the idle cycles of the functional units can be derived as:

$$MUL1, MUL2 : s4$$
$$ADD1 : s2, s3$$
$$SUB1 : s1, s2$$
$$CMP1 : s1, s3, s4$$

For functional units that have one or more idle controller states, it is possible to insert transparent latches at the functional unit's inputs to perform operand isolation, as shown in Figure 4.13. The latch enable signals for the latches at the inputs of a functional unit can be derived directly from its idle controller states. The expressions for the latch enable signals $LE1, \ldots, LE4$ in Figure 4.13 are:

$$LE1 = LE3 = x4$$
$$LE2 = x1 + x2$$
$$LE4 = x2 + x3$$

Latches have not been inserted at the inputs of $CMP1$ because the values in the registers that feed its inputs do not change in the cycles in which it is idle. The above observation leads to an alternative method to eliminate unnecessary activity in functional units – avoiding excessive register sharing during high-level

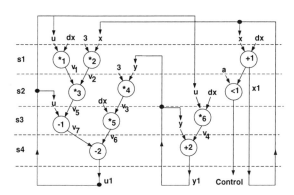

Figure 4.12. Operand isolation during high-level synthesis: Scheduled DFG

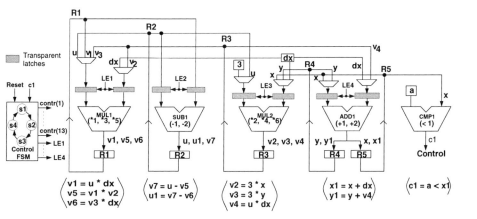

Figure 4.13. Operand isolation during high-level synthesis: RTL circuit

synthesis, or register duplication as a post-synthesis process. However, the extra power consumed by additional registers may be larger than the power overhead incurred by inserting transparent latches at each functional unit's inputs.

102   HIGH-LEVEL POWER ANALYSIS AND OPTIMIZATION

4.5 POWER MANAGEMENT THROUGH CONSTRAINED REGISTER SHARING

The operand isolation technique presented in the previous subsection attempts to eliminate spurious activity at the inputs of embedded resources (*e.g.* functional units) by inserting transparent latches into the RTL implementation. Apart from incurring power and area overheads due to the addition of extra circuitry, operand isolation also requires some delay constraints (the disabling transition at the transparent latch enable input should arrive before its data input can change). Satisfaction of the delay constraints may require the addition of extra circuit delay in the critical path, which may not be acceptable for high-performance designs.

This section demonstrates that the manner in which register allocation and variable binding (the assignment of variables to registers) are done can have a profound impact on spurious switching activity in functional units of the RTL implementation. Thus, an alternative to performing operand isolation after high-level synthesis is to perform judicious assignment of variables to registers to minimize or ensure the absence of spurious operations. Techniques to perform constrained register sharing to significantly reduce or eliminate spurious switching activity in the functional units of the RTL implementation are presented in [106]. While constrained register sharing may, in general, lead to an increase in the number of registers, in practice this overhead was observed to be small, and sometimes zero, for several designs [106].

Example 4.5 *Consider the scheduled DFG shown in Figure 4.14. Each operation in the DFG is annotated with its name (placed inside the circle representing the operation) and the name of the functional unit instance it is mapped to (placed outside the circle representing the operation). Each variable in the DFG is annotated with its name. Clock cycle boundaries are denoted by dotted lines. The schedule has five control steps, $s1, \ldots, s5$. Control step $s5$ is used to hold the output values in the registers and communicate them to the environment that the*

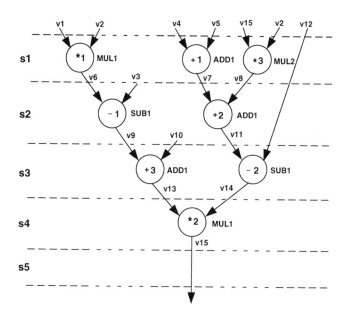

Figure 4.14. A scheduled CDFG to illustrate execution of spurious operations

design interacts with, and to load the input values into their respective registers for the next iteration.

In order to assess the impact of variable assignment on power consumption, consider two candidate assignments, Assignment 1 and Assignment 2, shown in Table 4.1. The architectures obtained using these assignments were subject to logic synthesis optimizations, and placed and routed using a 1.2 $micron$ standard cell library. The transistor-level netlists extracted from the layouts were simulated using a switch-level simulator with typical input traces to measure power. For the circuit Design 1, synthesized from Assignment 1, the power consumption was 30.71mW, and for the circuit Design 2, synthesized from Assignment 2, the power consumption was 18.96mW.

104   HIGH-LEVEL POWER ANALYSIS AND OPTIMIZATION

Table 4.1.   Two variable assignments for the scheduled DFG shown in Figure 4.14

Register	Assignment 1	Assignment 2
R1	v1, v7, v11, v13	v1, v13
R2	v2, v8, v10, v14	v2
R3	v3, v5, v9	v4, v8, v10
R4	v4, v6	v5, v7, v9
R5	v12	v12
R6	v15	v3
R7	-	v6, v11
R8	-	v14
R9	-	v15

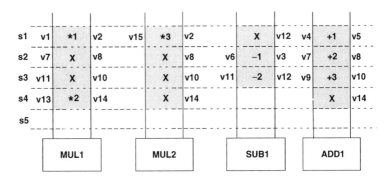

Figure 4.15.   Switching activity in the functional units of *Design 1*

To explain the significant difference in power consumption of the two designs, let us analyze the switching activities in the functional units that constitute Design *1 and* Design 2 *using Figures 4.15 and 4.16, respectively. In these figures, each functional unit is represented by a labeled box. The vertical lines which feed the box represent a duration equaling one iteration of execution of the DFG. Each control step is annotated with (i) the symbolic values that appear at the inputs of the*

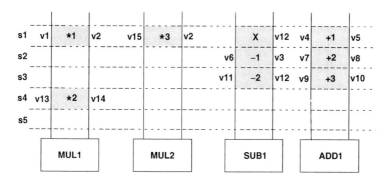

Figure 4.16. Switching activity in the functional units of *Design 2*

functional unit in the implementation, and (ii) the operation, if any, corresponding to the computation performed by the functional unit. For example, for functional unit $MUL1$ in Figure 4.15 in control step $s1$, variables $v1$ and $v2$ appear at its left and right inputs, respectively. The computation performed by $MUL1$ in control step $s1$ corresponds to operation $*1$ of the DFG. Cycles in which one or both inputs of a functional unit change, causing power dissipation, are shaded in the figures. Each variable change can be associated with the execution of a new operation. Cycles during which spurious input transitions, which do not correspond to any DFG operations, occur are marked with an X. The power consumption associated with these operations can be eliminated without altering the functionality of the design. A functional unit does not perform a spurious operation when both its inputs are unaltered. A cycle in which an input of a functional unit does not change is not marked with any variable. For example, in Figure 4.16, the inputs of functional unit $MUL1$ do not change from control step $s1$ to control step $s2$. Therefore, the inputs of $MUL1$ are unmarked in control step $s2$. An inspection of Figures 4.15 and 4.16 reveals that, while the functional units that constitute Design 1 execute seven spurious operations, the functional units that constitute Design 2 execute only one spurious operation. This explains the difference in power consumption between the two designs.

106 HIGH-LEVEL POWER ANALYSIS AND OPTIMIZATION

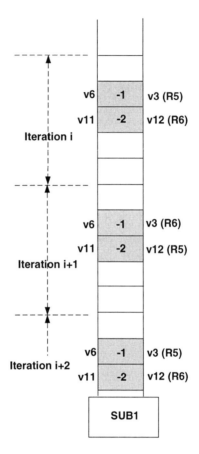

Figure 4.17. Eliminating spurious operations using dynamic variable rebinding

Example 4.6 *The previous example illustrated that constrained register sharing can significantly reduce the number of spurious operations in a circuit. This example illustrates a technique, called dynamic variable rebinding which, in combination with variable assignment, can completely eliminate spurious operations [106]. Consider functional unit $SUB1$ in* Design 2. *An inspection of Figure 4.16 reveals that it executes a spurious operation in control step $s1$ of every iteration. This* is because the multiplexer at its input selects register $R5$, to which variable $v12$ is assigned, from control step $s3$ of each iteration to control step $s1$ of the next iteration. Since $v12$ acquires a new value in control step $s1$ of each iteration, $SUB1$ computes $v11 - v12$ in control step $s1$, which is spurious, since the value of $v11$ corresponds to the previous iteration while the value of $v12$ corresponds to the current iteration. This problem would persist even if $v3$ were selected at the input of $SUB1$, instead of $v12$. This is because $v3$ is generated only at the end of control step $s1$, causing $SUB1$ to evaluate $v11 - v3$ with the old value of $v3$. Note that, in order to avoid the spurious operation, it is necessary to preserve the old value of $v12$ (from the previous iteration) at the input of $SUB1$ until the new value of $v3$ in the current iteration is born.

In this case, the spurious operation can be eliminated, without paying a price in terms of the number of registers used, by swapping the variables assigned to registers $R5$ and $R6$ in alternate iterations. In even iterations, $v3$ is mapped to $R6$ and $v12$ to $R5$, and vice versa in odd iterations. Figure 4.17 illustrates this transformation over three iterations of execution of the design. In this figure, the line representing the right input of $SUB1$ is annotated with the register containing the selected variable. As shown in the figure, register $R6$ stores variable $v12$ in iteration i, $v3$ in iteration $i + 1$, and so on. Since variable $v3$ is stored in register $R6$ in iteration $i + 1$, the old value of $v12$ is preserved at the input of $SUB1$ until $v3$ is born, thus avoiding the spurious operation marked X, shown in Figure 4.16.

4.6 CONTROLLER-BASED POWER MANAGEMENT

The operand isolation techniques presented in the previous section involve the addition of extra circuitry, including transparent latches and possibly some logic to generate their enable signals. This section presents a low-overhead power management technique that does not require the addition of any extra circuitry like transparent latches. It is based on minimally re-designing the existing control logic in order to reconfigure the multiplexer networks and functional units in the data path to minimize unnecessary switching activity [107]. Another distinction of this technique is that while conventional power management techniques seek to completely eliminate activity in the targeted sub-circuit, the controller-based technique may significantly reduce, but may not completely eliminate, such activity. Thus, it is frequently possible to reduce activity while avoiding the overheads associated with conventional techniques. Controller-based power management is well-suited to control-flow intensive designs, which pose challenges to other techniques such as operand isolation due to the following characteristics:

- Power consumption is dominated by an abundance of smaller components like multiplexers, while functional units may account for a small part of the total power [108]. The power overheads due to the insertion of transparent latches is comparable to the power savings obtained when power management is applied to sub-circuits such as multiplexer networks.

- The signals that detect idle conditions for various sub-circuits are typically late-arriving (for example, due to the presence of nested conditionals within each controller state, the idle conditions may depend on outputs of comparators from the data path). As a result, the timing constraints which must be imposed in order to apply conventional power management techniques (the enable signal to the transparent latches must settle before its data inputs can change) are often not met.

- The presence of significant glitching activity at control as well as data path signals needs to be accounted for in order to obtain maximal power savings.

The following example illustrates the effect of re-specifying control signals on the activity of data path signals. Consider an RTL circuit that implements the send process of the X.25 telecommunication protocol [59] that is shown in Figure 4.18. Figure 4.19(a) shows an extracted part of the X.25 data path that consists of an ALU and the multiplexer trees that feed it. Figure 4.19(b) shows (i) the logic expressions for the control signals that feed the ALU and its multiplexer trees (as before, x

POWER MANAGEMENT   109

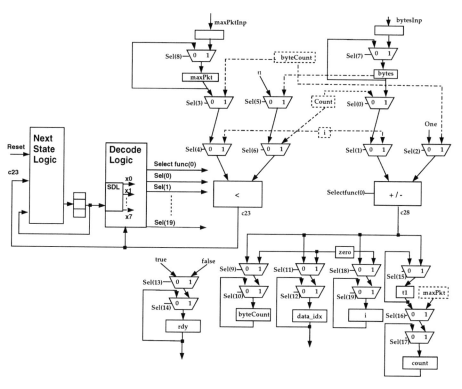

Figure 4.18.   RTL circuit implementing the send process of the X.25 protocol

represents a decoded state variable, *i.e.* $xi = 1$ when the controller is in state si), and (ii) an *activity graph* for the ALU, that indicates the operations performed by the ALU in each controller state. The vertices and arcs in the activity graph correspond to the controller states and state transitions. Each vertex in the activity graph is labeled with the computation performed by the ALU in the corresponding controller state. For example, consider controller state $s2$. Control signals $Sel(0)$, $Sel(1)$, $Sel(2)$ and $Selectfunc(0)$ (function select input, $1 \Rightarrow +, 0 \Rightarrow -$) assume the logic values 1, 0, 1, and 0, respectively. From these values, it can be easily seen that the ALU performs the operation $bytes - byteCount$ in state $s2$.

110   HIGH-LEVEL POWER ANALYSIS AND OPTIMIZATION

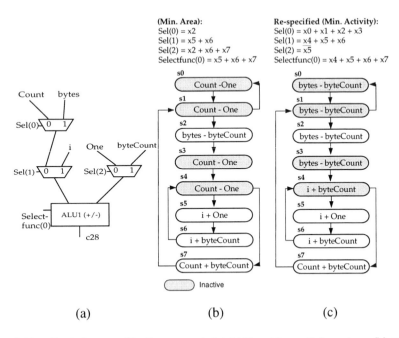

Figure 4.19. Control re-specification example (a) ALU and its multiplexer tree, (b) original control expressions and activity graph, and (c) re-specified control expressions and activity graph

In some states, like $s0$, the ALU may not be required to perform any operation, i.e. its result may be unused. Using the scheduling and assignment information from high-level synthesis, it is possible to easily identify such idle states based on the absence of operations assigned to the ALU. In Figure 4.19(b), idle states are indicated by shaded vertices. The computation performed by the ALU in idle states can be changed without affecting the functionality of the design. The control signals $Sel(0), Sel(1), Sel(2)$, and $Selectfunc(0)$ were re-specified using the techniques presented later in this section. The resulting logic expressions for the re-specified control signals and the modified activity graph for the ALU are shown in Figure 4.19(c). Note that the labels of the vertices in the activity graph

POWER MANAGEMENT   111

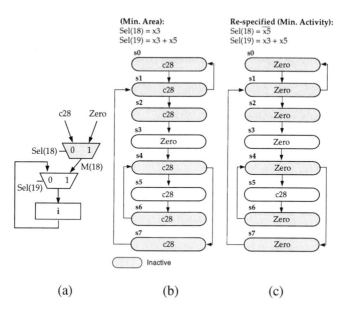

(a)   (b)   (c)

Figure 4.20. (a) Multiplexer tree feeding a register, (b) original control expressions and activity graph for signal M(18), and (c) re-specified control expressions and activity graph

have changed. Consider two consecutive cycles in the operation of the X.25 circuit, during which the controller makes the state transition $s6 \rightarrow s4$. Under the original control expressions, there is switching activity in the ALU since its operands change from i (left operand), $byteCount$ (right operand), and 1 (function select input) in $s6$ to $Count$, One, and 0 respectively in $s4$. Under the re-specified control expressions, however, all input operands to the ALU remain stable. Hence, we conclude that re-specification of the control signals feeding the ALU, or equivalently, re-labeling of the ALU's activity graph, can affect the switching activity, and hence power consumption, in the ALU. During the actual operation of the circuit, controller state transitions other than $s6 \rightarrow s4$ will also occur. In general, it is necessary to consider all incoming and outgoing arcs while deciding on how to (re-)label an idle vertex in the activity graph. We formalize

112   HIGH-LEVEL POWER ANALYSIS AND OPTIMIZATION

the use of state transition probabilities or transition counts in labeling the activity graph later on in this section.

Re-specifying the controller can also lead to reduced activity within multiplexer trees. Figure 4.20(a) shows another part of the X.25 data path that consists of the register that stores variable i together with the multiplexers feeding it. The original expressions for the control signals $Sel(18)$ and $Sel(19)$ are given in Figure 4.20(b). The activity graph for signal $M(18)$ (the output of the shaded 2-to-1 multiplexer), that indicates which operand ($c28$ or $Zero$) is selected at $M(18)$, is also shown in Figure 4.20(b). The shaded vertices in the activity graph correspond to states when the value of signal $M(18)$ is not used. Consider controller state transition $s7 \rightarrow s1$. Since the computation performed by the ALU changes from $count + byteCount$ in $s7$ to $bytes - byteCount$ in $s1$ (see Figure 4.19(c)), the value of the operand ($c28$) itself changes. Re-specifying the control signals as shown in Figure 4.20(c) eliminates this unnecessary activity at signal $M(18)$.

Re-labeling activity graphs using state transition counts and activity matrices

When deciding how to label an idle vertex in an activity graph, it is necessary to keep in mind the following issues.

- Different incoming and outgoing transitions into the idle state have different execution probabilities, and

- The values of data operands fed to multiplexer trees may themselves change. Hence, merely selecting the same operand does not ensure that switching activity is minimized.

Consider a less-than ($<$) comparator and its partial activity graph shown in Figure 4.21. Vertex $s3$ in the activity graph corresponds to an idle state for the comparator. The activity graph shows all vertices that have incoming arcs from or outgoing arcs to vertex $s3$. Vertices $s1$, $s2$, $s4$, and $s5$ have the labels $L1$ ($a < b$), $L2$ ($c < d$), $L2$, and $L3$ ($e < d$), respectively. We wish to assign one of the labels from the set $\{L1, L2, L3\}$ to vertex $s3$ such that the activity at the inputs of the comparator, and hence its power consumption, is minimized. While this discussion assumes that minimizing the average switching activity at the inputs of a block minimizes its power consumption, it is also possible to incorporate more sophisticated models for data path power consumption to drive controller re-specification.

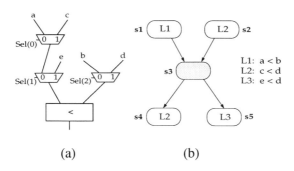

Figure 4.21. (a) Comparator and its multiplexer trees, and (b) activity graph used for re-labeling

For each arc $si \rightarrow sj$ in the activity graph, $P(si \rightarrow sj)$ is the probability of a controller state transition from si to sj. An *activity matrix* $AM_{si \rightarrow sj}$ for arc $si \rightarrow sj$ in the activity graph stores the costs in terms of average bit transitions incurred for various combinations of labels that can be assigned to vertices si and sj. The rows (columns) of $AM_{si \rightarrow sj}$ are indexed by the set of possible labels that can be assigned to si (sj). For example, consider the arc from vertex $s1$ to vertex $s3$ in the activity graph shown in Figure 4.21. Since the label of $s1$ is fixed to $L1$, the activity matrix $AM_{s1 \rightarrow s3}$ has valid entries in only one row, that is indexed by $L1$. We can (re-)label the vertices in the activity graph that correspond to idle states while attempting to minimize the *Labeling Cost* that is given by:

$$Labeling\ Cost = \sum_{all\, si \rightarrow sj} AM_{si \rightarrow sj}[L(si), L(sj)].P(si \rightarrow sj)$$

where $L(si)$ refers to the label assigned to vertex si in the activity graph. For the example of Figure 4.21, the cost of labeling vertex $s3$ with a label L^* is given by:

$$AM_{s1 \rightarrow s3}[L1, L^*].P(s1 \rightarrow s3) + AM_{s2 \rightarrow s3}[L2, L^*].P(s2 \rightarrow s3)\ +$$

$$AM_{s3\to s4}[L^*, L2].P(s3 \to s4) + AM_{s3\to s5}[L^*, L3].P(s3 \to s5)$$

The objective, therefore, is to choose $L^* \in \{L1, L2, L3\}$ such that the labeling cost, as given by the above equation, is minimized.

Algorithms for performing controller-based power management are described in detail in [107]. While controller re-specification seeks to minimize zero-delay switching activity at the inputs of idle blocks, it may have several negative side-effects, including:

- Increase in the circuit delay, due to the formation of long paths.

- Increase in glitching power consumption, due to increased glitching activity at control and/or data signals, which may offset the power savings.

- Formation of false combinational loops in the circuit, which may be unacceptable due to the limitations imposed by lower-level design tools.

The techniques presented in [107] judiciously perform controller-based power management, while avoiding the above negative effects.

4.7 CONCLUSIONS

Power management is one of the most popular power management techniques employed in low power design methodologies due to its potential for delivering large and unambiguous power savings, and since it requires minimal intrusion into the other steps of the design flow. Hence, it is not unreasonable to expect that a significant focus of future commercial tools for high-level power optimization will be on power management. While several techniques for applying power management to a given architecture exist, the effect of system-level trade-offs and architecture synthesis or selection on power management opportunities is still not well understood, and is an area that requires further investigation.

5 HIGH-LEVEL SYNTHESIS FOR LOW POWER

High-level synthesis (also called behavioral synthesis or architectural synthesis) refers to the process of transforming a functional or behavioral specification of a design into a structural RTL implementation. A typical high-level synthesis process involves several subtasks including behavioral transformations, module selection, clock period selection, scheduling, and resource sharing, and RTL circuit generation. High-level synthesis has a large impact on power consumption, which, if properly exploited, can lead to large power savings. This chapter analyzes the effect of various high-level synthesis subtasks on power, and presents various techniques that can be used to optimize power consumption during high-level synthesis.

5.1 BEHAVIORAL TRANSFORMATIONS

Behavioral transformations refer to changes to the computational structure of an algorithm that preserve its input-output behavior, while resulting in an implementation that is optimized for one or more design metrics. Several high-level synthesis systems have incorporated comprehensive sets of transformations for optimization of area and performance [109, 110, 111, 112, 113]. Examples of transformations that can be applied to behavioral descriptions, or CDFGs, include:

- Algebraic transformations such as associativity, commutativity, and distributivity.

- Common subexpression elimination.

- Constant propagation and dead code elimination.

- Strength reduction, such as replacement of constant multiplications with shift-and-add operations.

- Numerical transformations, such as word-length reduction.

- Loop transformations, such as loop unrolling, winding, re-ordering, and merging.

- Behavioral retiming and pipelining.

- Hierarchy transformations, such as procedure inlining and extraction.

Behavioral transformations have been shown to have a very significant impact on the area and performance of the implementation. Optimizing power using transformations has been studied more recently [3, 4, 5, 114].

5.1.1 Enabling supply voltage reduction using transformations

The quadratic influence of the supply voltage on power dissipation makes supply voltage scaling an attractive approach for power reduction, as described in Chapter 2. Reducing the supply voltage alone, however, has a negative impact on the circuit delay, making it difficult to apply this methodology to high-performance systems. The negative impact on delay can be overcome by using technology optimizations such as threshold voltage reduction, multiple threshold voltages, *etc.* [115]. However, other issues such as standby power and noise margins impose limitations on the extent to which such technology optimizations can be performed.

Another approach for recovering the delay degradation due to supply voltage reduction is to use architectural transformations that have conventionally been shown to be very effective in performance optimization [31]. A framework for the use of such transformations in the context of data-flow intensive applications is presented in [3]. Given a behavioral specification and a throughput constraint, the idea is to apply performance optimizing transformations and transformation sequences that enhance the throughput of the design as much as possible (more than required by the constraint), and use the slack thus generated to reduce the supply voltage until the transformed design just meets the original throughput constraint.

Example 5.1 *Consider a first-order Infinite Impulse Response (IIR) filter, shown in Figure 5.1(a), that is described by the following difference equation.*

$$Y_N = K * Y_{N-1} + X_N \tag{5.1}$$

Assuming that the addition and constant multiplication operations take one clock cycle each, the critical path for the DFG of Figure 5.1(a) has two clock cycles. Suppose that we can implement the DFG structure of Figure 5.1(a) such that it just meets the required sample period[1] *at a supply voltage of 5V. The normalized*

---

[1] Sample period is the time interval between the application of two consecutive input samples to the design. The sample period is the inverse of throughput.

118 HIGH-LEVEL POWER ANALYSIS AND OPTIMIZATION

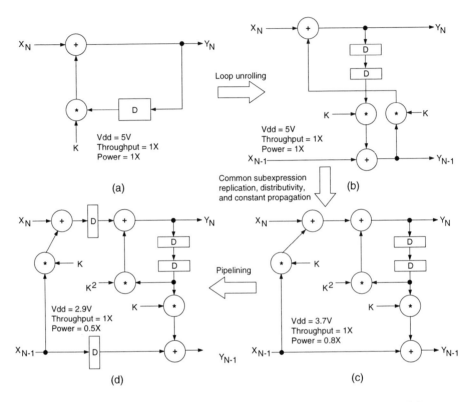

Figure 5.1. Using transformations to enable supply voltage reduction [3]

throughput and power are indicated in Figure 5.1(a) as $1X$ *and* $1X$, *respectively. None of the simple transformations is applicable directly to the DFG shown in Figure 5.1(a). Applying loop unrolling results in the DFG shown in Figure 5.1(b), which has the same performance and power characteristics. However, loop unrolling enables us to apply other transformations, such as common subexpression replication, distributivity, and constant propagation, resulting in the DFG shown in Figure 5.1(c). The critical path for the DFG of Figure 5.1(c) has three clock cycles for computing two output samples, which is less than the original critical* path of two clock cycles per output sample. Exploiting this performance slack, the supply voltage can be reduced to 3.7V while maintaining the throughput of the original design. However, the switched capacitance per sample period for the DFG in Figure 5.1(c) is higher than the original design, due to an increase in the number of operations. In effect, the normalized power of the implementation of the DFG shown in Figure 5.1(c) is $0.8X$. Applying pipelining to the DFG of Figure 5.1(c) results in the DFG of Figure 5.1(d), which has a critical path of two cycles for the computation of two output samples. As a result, the supply voltage can be further reduced to 2.9V, resulting in a normalized power of $0.5X$ for the implementation of the DFG shown in Figure 5.1(c) at the same performance as the original design.

5.1.2 Minimizing switched capacitance

It is also possible to reduce switched capacitance using one or more of the following behavioral transformations.

- Reducing the number of operations required.
- Replacing high power consuming operations (such as multiplications) with less power consuming operations (such as shift-and-add operations).
- Architecture re-structuring to reduce zero-delay activity at intermediate variables.
- Architectural path balancing to reduce glitching activity.
- Word-length reduction.

The application of transformations to reduce switched capacitance is illustrated through the following examples.

Example 5.2 *Consider the computation of the expression $X^2 + A.X + B$. The straightforward DFG that performs the computation, which is shown in Figure 5.2(a), requires two multiplications and two additions, and has a critical*

120   HIGH-LEVEL POWER ANALYSIS AND OPTIMIZATION

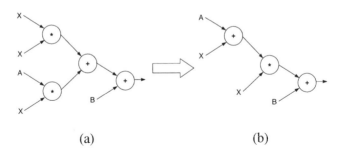

Figure 5.2. Minimizing switched capacitance by reducing the number of operations in the DFG [3]

path of three clock cycles assuming each operation requires one clock cycle. By applying the distributivity transformation, the expression can be re-written as $(A + X).X + B$, and is represented by the DFG shown in Figure 5.2(b). The transformed DFG requires one less multiplication than the original DFG, and has the same critical path, leading to a significant savings in power consumption.

Example 5.3 *Consider the DFG for complex number multiplication that is shown in Figure 5.3(a). X_r and X_i represent the real and imaginary components of the complex number X. A is a constant complex number whose real and imaginary components are A_r and A_i, respectively. The original DFG contains four multiplications, one addition, and one subtraction, and requires two clock cycles assuming each operation requires one clock cycle to execute. The application of distributivity, common subexpression elimination, and constant propagation in an appropriate sequence, results in the DFG shown in Figure 5.3 that contains one more addition, but one less multiplication. Hence, a multiplication has been replaced with an addition, which is less power consuming.*

The following example shows how the computation structure of an algorithm can be re-organized so as to minimize the word length of some of the variables and operations involved, leading to a reduction in power consumption.

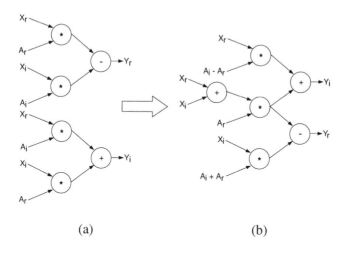

Figure 5.3. Minimizing switched capacitance by strength reduction [3]

Example 5.4 *Consider an N-tap Finite Impulse Response (FIR) filter, that computes the following expression.*

$$Y_j = \sum_{k=0}^{N-1} \{P_k\}_j \tag{5.2}$$
$$\{P_k\}_j = C_k . X_{j-k}$$

where C_k are constants that represent the coefficients of the filter, X_j represents the input value at discrete time instant j, and Y_j represents the output value at time j. The DFG for the straightforward implementation of the above equation, called the direct form, is shown in Figure 5.4(a).

Consider the expression for Y_{j+1} that is given by the following equation.

$$Y_{j+1} = \sum_{k=0}^{N-1} \{P_k\}_{j+1}$$
$$\{P_k\}_{j+1} = C_k . X_{j-k+1}$$

122 HIGH-LEVEL POWER ANALYSIS AND OPTIMIZATION

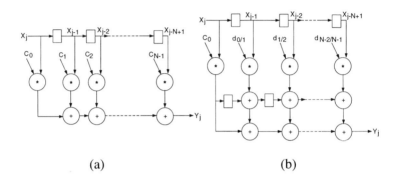

Figure 5.4. Using differential coefficients to minimize word-length of multiplication operations [4]

Each product term, $\{P_k\}_{j+1}$, except the first one ($C_0.X_{j+1}$), can be re-written as follows.

$$C_k.X_{j-k+1} = C_{k-1}.X_{j-k+1} + (C_k - C_{k-1}).X_{j-k+1} \quad for\ k = 1\ to\ N - 1$$

The second term ($C_k - C_{k-1}$) is the first-order difference between consecutive coefficients, and is represented as $d_{k-1/k}$. In addition, note that the first term in the above equation can be expressed in terms of the product terms involved in the computation of the previous output sample, Y_j, as shown below.

$$\begin{aligned} C_{k-1}.X_{j-k+1},\ for\ k = 1\ to\ N-1, &= C_k.X_{j-k},\ for\ k = 0\ to\ N-2, \\ &= \{P_k\}_j,\ for\ k = 0\ to\ N-2 \end{aligned}$$

Thus, the first $N - 1$ product terms that are used to compute Y_j can be stored and re-used to compute Y_{j+1}, along with the first-order difference terms.

The DFG for the implementation of the FIR filter that uses the first-order difference terms is shown in Figure 5.4(b). Note that the implementation of Figure 5.4(b) requires the same number of multiplications, N more additions, and N more delay (storage) units, as compared to the direct form implementation of

Figure 5.4(a). However, it is often the case in practice that the magnitudes of the differences between successive coefficients are much smaller that the magnitudes of the coefficients themselves. As a result, the word lengths of the constant operands in the multiplications of Figure 5.4(b) are significantly smaller than those in Figure 5.4(a). This leads to a significant decrease in the power consumed to compute the multiplications, which often outweighs the aforementioned overheads, leading to power savings [4]. The above technique can be generalized to use higher-order differences to implement FIR filter computations, as described in [4].

The previous examples in this subsection explored the effects of behavioral transformations on the number and types of operations required to perform the given computation. Behavioral transformations also impact the signal statistics of the intermediate variables involved in the computation, which can be exploited to minimize switching activity and hence reduce power. The application of commutativity and associativity transformations to minimize the switching activity of the variables in linear DSP circuits is presented in [5]. For parallel implementations in which no resource sharing is performed, a reduction in switching activity in DFG variables directly translates to a reduction in switching activity in the registers and inputs of functional units in the RTL implementation. While parallel implementations of DFGs may, in general, be inefficient, many linear DSP circuits are implemented without resource sharing in order to use bit-serial implementations or to enable highly pipelined designs.

In order to explore the effect of transforming the DFG structure, let us study the average activities at intermediate variables generated by the two common types of operations present in linear DSP circuits – constant multiplication and addition operations. Consider a constant multiplier, which multiplies an m-bit data value X by an m-bit constant A (X and A are represented in two's complement form). Applying random input sequences at X and observing the average switching activity per bit at the multiplier output for various values of A ranging from 0.0 to 1.0 results in the graph shown in Figure 5.5. As expected, when the constant value is 0, there is no activity at the output of the multiplier. As the magnitude of

124   HIGH-LEVEL POWER ANALYSIS AND OPTIMIZATION

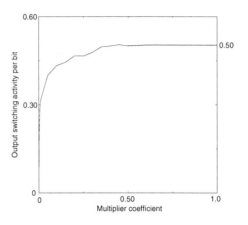

Figure 5.5. Average switching activity at the output of a constant multiplier *vs.* constant value [5]

the constant increases, the activity at the multiplier output increases, and becomes equal to the switching activity at X when the constant becomes 1.0. A similar trend is observed as the value of the constant is decreased from 0 to -1.0.

Next, consider an adder which adds two m-bit data values X_1 and X_2. Assuming that X_1 and X_2 are random and uncorrelated, the variation of the average switching activity ($Sw\_act(Y)$) at the adder output Y as a function of the input switching activities ($Sw\_act(X_1)$ and $Sw\_act(X_2)$) can be shown to closely obey the following relationship [5].

$$Sw\_act(Y) = \max(Sw\_act(X_1), SW\_act(X_2)) \quad (5.3)$$

Example 5.5 *Consider the computation of a weighted linear expression, that forms the basis of all linear time-invariant signal processing systems.*

$$Y = \sum_{i=1}^{n} A_i.X_i \quad (5.4)$$

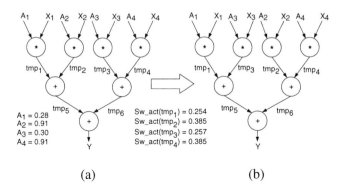

Figure 5.6. Minimizing switching activity using transformations

In particular, let us consider the case where $n = 4$. One possible DFG for such a computation is shown in Figure 5.6(a). The DFG is annotated with the values of the various constants, and the average switching activities per bit at the variables that represent the results of the constant multiplications. Utilizing the observation described earlier about the activities at the outputs of adders, we can conclude that the switching activities at the remaining variables in the DFG are as follows:

$$\begin{aligned} Sw\_act(tmp_5) &= \max(Sw\_act(tmp_1), Sw\_act(tmp_2)) = 0.385 \\ Sw\_act(tmp_6) &= \max(Sw\_act(tmp_3), Sw\_act(tmp_4)) = 0.385 \\ Sw\_act(Y) &= \max(Sw\_act(tmp_5), Sw\_act(tmp_6)) = 0.385 \end{aligned}$$

(5.5)

Now, consider the transformed DFG structure shown in Figure 5.6(b). The computation tree has been re-structured so that the constants of the constant multipliers are in increasing order from left to right in the figure. We compute the activities at the remaining variables of the DFG as follows.

$$\begin{aligned} Sw\_act(tmp_5) &= \max(Sw\_act(tmp_1), Sw\_act(tmp_3)) = 0.257 \\ Sw\_act(tmp_6) &= \max(Sw\_act(tmp_2), Sw\_act(tmp_4)) = 0.385 \end{aligned}$$

$$Sw\_act(Y) = \max(Sw\_act(tmp_5), Sw\_act(tmp_6)) = 0.385 \tag{5.6}$$

It can be seen that the switching activity at intermediate variable tmp_5 is smaller in case of the DFG of Figure 5.6(b) than in case of the DFG of Figure 5.6(a), and the switching activities of all other variables are equal.

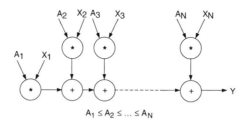

Figure 5.7. Activity reduction in a linear chain [5]

The observations of the above example can be generalized as follows.

- Consider a balanced computation tree with n inputs that computes the expression shown in Equation (5.4). The average switching activity over all intermediate variables is minimized if $A_1 \leq A_2 \ldots \leq A_n$ or $A_1 \geq A_2 \ldots \geq A_n$.

- Consider a serial chain, shown in Figure 5.7 that computes the expression shown in Equation (5.4). The average switching activity over all intermediate variables is minimized if $A_1 \leq A_2 \ldots \leq A_n$.

5.2 MODULE SELECTION

Module selection refers to the process of mapping operations from the CDFG to component templates from the RTL library. Note that only a functional unit template, and not a specific instance, is associated with each operation. For example, an addition operation may be implemented using a ripple-carry adder, carry-lookahead adder, carry-select adder, *etc.* Each distinct implementation for the same operation may have different area, delay, and power characteristics [116, 117]. For example, a ripple-carry adder is slower but more switched capacitance efficient, while a carry-lookahead adder is faster but incurs higher switched capacitance. Similar trade-offs exist for other functional units and RTL components. These trade-offs could be exploited during the module selection process, by considering power together with area and delay constraints as co-objectives, as shown by the following example.

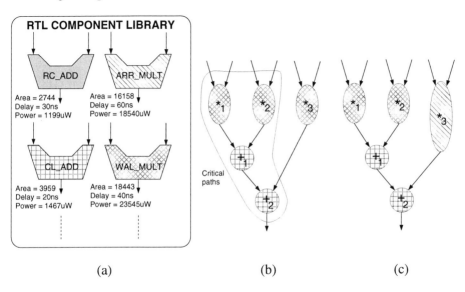

Figure 5.8. Minimizing power consumption through module selection

Example 5.6 *Consider the RTL component library shown in Figure 5.8(a), and the DFG shown in Figure 5.8(b). Each operation in the DFG of Figure 5.8(b) has been mapped to the fastest functional unit template available in the library as indicated by the shading, in order to meet the specified performance constraint of* $85ns$. *However, not all the operations of the DFG need to be mapped to the fastest* *possible functional units. In particular, it is sufficient to map operations on critical paths in the DFG, which are indicated in Figure 5.8(b), to fast functional units. The slack available for the off-critical-path operations can be exploited to use slower functional units, which also tend to be more switched capacitance efficient. Figure 5.8(c) shows such a mapping, where operation $*_3$ has been mapped to a slower multiplier template, that consumes less energy per multiplication.*

The availability of a diverse module library, in which several implementation choices exist for each RTL component, is critical in obtaining maximal power reductions. Techniques and algorithms to perform module selection to minimize power consumption are provided in [28, 29, 30]. It should be noted that while the above example illustrated the effect of module selection on switched capacitance, module selection also affects opportunities for supply voltage scaling. The use of faster functional units may lead to an implementation that exceeds the required performance, and the resulting slack can be used to reduce power by scaling the supply voltage.

Using multiple supply voltages during high-level synthesis

The use of multiple supply voltages is a promising technique to obtain low power implementations at reduced performance overheads. This idea has been exploited at the logic level by identifying off-critical-path sub-circuits that can be operated at a lower supply voltage [118]. In the context of high-level synthesis, one way to utilize multiple supply voltages is to have an RTL component library that contains multiple versions of each component corresponding to different supply voltages. Voltage level converters may be required to communicate between logic blocks that operate at different voltage levels. The module selection process can be extended to assign each CDFG operation to a library component template and a specific supply voltage, and insert the necessary level converters during the process. The delay models used during scheduling need to be sensitive to the dependence of delay on supply voltage, and the resource sharing process needs to be constrained to disallow sharing of a functional unit for operations that are assigned to different supply voltages.

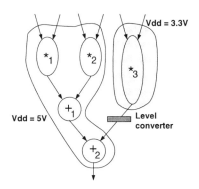

Figure 5.9. Using multiple supply voltages to minimize power

Example 5.7 *Consider the DFG shown in Figure 5.9. The off-critical-path operation, $*_3$, has been assigned to a functional unit template that operates at a lower supply voltage (3.3V) than the rest of the DFG. A level conversion operation is inserted into the DFG to convert the result of $*_3$ to the higher logic level. The level conversion operation can either be implemented using a separate level converter circuit such as a differential cascode voltage switch (DCVS) gate, or can be integrated into a register to reduce its overhead [118].*

Various algorithms to perform supply voltage assignment and level converter insertion together with module selection and scheduling, to minimize power given performance constraints, are presented in [29, 33, 34, 35].

5.3 RESOURCE SHARING

Resource sharing refers to the process of mapping the operations and variables in the CDFG to specific structural entities such as functional units and registers,

130   HIGH-LEVEL POWER ANALYSIS AND OPTIMIZATION

and defining interconnection among the functional units and registers, to form the RTL implementation. Since the resource sharing process provides the mapping from a design's functionality to its structure, it directly impacts the power consumption by determining the switching activity at various signals, as well as the physical capacitance of the macroblocks, buses, and wires that constitute the implementation.

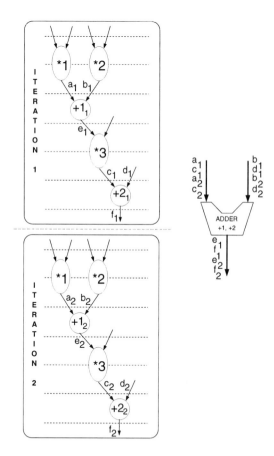

Figure 5.10. Effect of resource sharing on the switching activity in a shared resource [6]

5.3.1 Exploiting signal correlations to reduce switched capacitance

Resource sharing results in values of variables in the behavioral description being time-multiplexed in the registers of the RTL implementation. Similarly, the values that appear at input operands of operations are time-multiplexed to appear at the inputs of functional units, and the values required to be transferred between functional units and registers are sequenced to appear on the interconnect units, such as buses and multiplexers, that connect them. Hence, the word-level temporal correlations of the values that appear at various data path signals are determined by the correlations among variables and input operands of operations that are grouped together during resource sharing [6, 83, 102, 119, 120, 121]. These word-level temporal correlations in turn determine the bit-level switching activities at the signals in the RTL circuit, as explained in Chapter 3, and hence the power consumption in the various RTL circuit components.

Example 5.8 *Consider the DFG shown in Figure 5.10. In particular, let us focus on the two addition operations, $+_1$ and $+_2$ that share the same adder in the RTL implementation. Two consecutive iterations of the DFG are shown in Figure 5.10. The sequence of operations performed by the adder is $+1_1$, $+2_1$, $+1_2$, $+2_2$, ..., where the subscripts stand for the iteration number. Hence, the values seen at the adder inputs are (a_1, b_1), (c_1, d_1), (a_2, b_2), (c_2, d_2), ..., where a_i is the value taken on by variable a in the ith iteration of the CDFG, and so on. The switching activity at the adder inputs is determined by:*

- *Intra-iteration effects: The Hamming distance between the values of a_1 and c_1 (also b_1 and d_1) in the first iteration, and a_2 and c_2 (also b_2 and d_2) in the second iteration.*

- *Inter-iteration effects: The Hamming distance between the values of c_1 and a_2 (also d_1 and b_2).*

It is possible to exploit correlations between variables in the behavioral description to minimize the switched capacitance in the RTL implementation. Correlations may exist between variables in a behavioral description as a result of one or more of the following factors:

- Several applications, such as DSP applications are characterized by slow-varying inputs, *i.e.* the inputs are highly temporally self-correlated, leading to several internal variables being temporally correlated as well. These temporal correlations are typically inter-iteration correlations. In [3], it is pointed out that for such applications, an architecture with little or no hardware sharing may be better than a highly hardware shared one because hardware sharing could destroy these temporal correlations and create a lot of switching activity.

- The values of different variables in the same iteration may be correlated due to re-convergent fanout in the CDFG.

- The value assumed by an input in an iteration may be correlated with the value assumed by other inputs in the same iteration. For example, in high-quality audio applications, one might have different correlated sound tracks fed to different speakers to create a surround-sound effect.

- The functional relationship between the signals that is imposed by the computation may lead to correlation. For example, the variable that represents the result of an operation may be correlated with the variables that represent the input operands to the operation. This is illustrated in Table 5.1, which shows the bit-level correlation between one input operand and the output of the operation, for various types of operations and a bit-width of 4.

Table 5.1. Bit-level correlations between input and output values of operations

op	$+$	$*$	AND	OR
correlation	0.500	0.617	0.75	0.75

Various resource sharing algorithms that exploit the observations presented in this subsection to minimize switched capacitance, and hence power, are presented in [6, 83, 102, 119, 120, 121].

5.3.2 Exploiting regularity to minimize interconnect power

The previous subsection described the effect of resource sharing on switched capacitance in the functional units and registers of the RTL implementation. In addition, the process of resource sharing also maps the communication patterns of the CDFG into interconnections among functional units and registers through multiplexers and/or buses in the data path. Ignoring the interconnect power during resource sharing may lead to architectures that consume a lot of interconnect power. This effect is amplified as interconnect begins to dominate the area, delay, and power of circuits in deep submicron technologies.

Regularity in an algorithm refers to the repeated occurrence of computational patterns within it. One way to detect such computational patterns is using the notion of *E-templates* introduced in [7]. An *E-template* is a pattern that consists of a pair of CDFG nodes, characterized by their types, connected by an edge. Some examples of E-templates are: $mult \rightarrow add.left$ (a multiply operation that feeds the left input of an add operation), $mult \rightarrow add.right$, $add \rightarrow add.right$, etc. An instance of an *E-template* in a CDFG is called an *E-instance*.

Regularity can be exploited to reduce interconnect power by detecting instances of repetitive patterns in the computation, and performing resource sharing in a manner such that the same interconnect structure in the data path is re-used for as many instances of the computation patterns as possible. This leads to a reduction in the interconnect requirements of the architecture, as illustrated by the following example.

Example 5.9 *Consider the set of E-instances for part of a DFG that is shown in Figure 5.11. The figure shows two different assignments that have the same resource requirements with respect to the operations shown. In the case of Figure 5.11(a),*

134   HIGH-LEVEL POWER ANALYSIS AND OPTIMIZATION

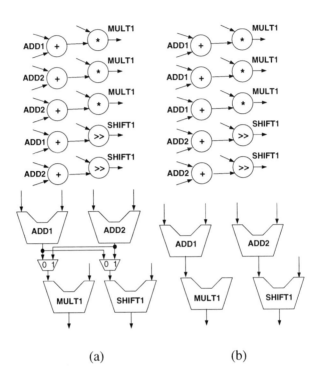

Figure 5.11. Exploiting regularity to minimize interconnect power: (a) non-regular assignment, and (b) regular assignment [7]

resource sharing was performed without regard to preserving the regularity of the algorithm. The output of each adder needs to communicate with the left input of the multiplier as well as the left input of the shifter, requiring multiplexers at the multiplier and shifter inputs. On the other hand, in the case of Figure 5.11(b), resource sharing was performed while trying to assign the operations of all the instances of the $add \rightarrow mult.left$ E-template to the same adder and multiplier. This results in fewer multiplexers and fanouts at the outputs of the adders, leading to a reduction in interconnect power.

A resource sharing algorithm to exploit regularity is presented in [7], that is based on assigning E-templates as a whole (as opposed to individual operations) in order to minimize the number and size of multiplexers and buses, and the fanouts of wires or nets in the RTL implementation.

5.4 SCHEDULING

Scheduling refers to the process of assigning operations in the behavioral description to control steps or controller states in which they execute. Scheduling determines the cycle-by-cycle behavior of the design. In addition, operations (variables) that are active in the same control step must be assigned to different functional units (registers). Hence, resource sharing possibilities depend on the results of the scheduling process. Scheduling techniques can be divided into two large classes – resource-constrained scheduling, where a fixed set of resources is utilized during scheduling, and time-constrained scheduling where some metric of the performance of the design, *e.g.* the number of clock cycles required to perform the computation, is fixed. In addition, scheduling techniques also require the specification of a value for the system clock period. The choice of clock period can significantly affect the results of scheduling. Multicycling refers to the use of multiple cycles to execute an operation when the delay of the functional unit template it is assigned to is larger than the clock period. Chaining refers to the complementary situation where multiple operations with data dependencies are scheduled in a single control step, because the clock period is large enough to permit us to do so. The determination of an appropriate clock period to use during scheduling is in itself an important high-level synthesis subtask [48].

The effect of scheduling on average power consumption is complex, and related to the other high-level synthesis subtasks, including module selection and resource sharing, as explained below.

- Scheduling determines the sequence in which the various operations of the behavioral description are performed, and also dictates which operations and variables can share the same functional units and registers. Thus, scheduling can be used to enable resource sharing for low power by ensuring that correlated variables and operations with correlated operands are appropriately sequenced so that they can share the same resources [30, 102, 119] (see Subsection 5.3.1).

- Scheduling can be performed so as to enable maximum resource sharing between operations that belong to instances of the same computational pattern, resulting in maximal exploitation of regularity during resource sharing (see Subsection 5.3.2).

- Scheduling can be used to distribute the slacks or mobilities of various operations in the DFG appropriately so that some operations may be performed using slower, more energy-efficient functional units. Thus, scheduling has an impact on the power trade-offs through module selection (see Section 5.2).

5.4.1 Effect of scheduling on peak power consumption

Scheduling determines the distribution of operations over time, and hence affects the profile of the power consumption in the implementation over time (control steps or clock cycles). As explained in Chapter 2, reducing peak power is important due to packaging, cooling, and reliability considerations. The following example illustrates the effect of scheduling on peak power.

Example 5.10 *Consider the scheduled DFG shown in Figure 5.12(a). Assuming a constant power per operation model for simplicity, the profile of the power consumption requirements of the functional units is shown in the bar graph of Figure 5.12(a). The peak power consumption occurs in the first clock cycle, s_1, due to the simultaneous computation of the three multiplication operations.*

Figure 5.12(b) shows an alternative schedule where the slack for multiply operation $*_3$ has been utilized to move it to the second control step, $s2$. The resulting power profile, which is also shown in Figure 5.12(b), indicates that the

HIGH-LEVEL SYNTHESIS FOR LOW POWER   137

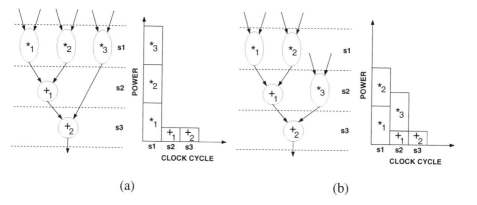

Figure 5.12.   Effect of scheduling on peak power consumption

peak power requirement of this schedule is significantly smaller than the peak power requirement of the schedule shown in Figure 5.12(a).

Scheduling techniques to minimize and satisfy constraints on the peak power consumption are described in [29]. Other high-level synthesis tasks, such as module selection and transformations also have an impact on peak power consumption. However, their effect has not yet been studied in detail.

5.4.2   Effect of clock period selection on power

Scheduling techniques can be employed with a target clock period for the implementation. The choice of the clock period used during scheduling affects power consumption directly and indirectly in the following ways.

- Larger values of the clock period lead to schedules with more functional unit chaining. Since the outputs of functional units can be highly glitchy, this leads to a significant increase in glitching power consumption.

- Larger values of the clock period may lead to a design that requires fewer clock cycles to process each input. This means that the clock distribution

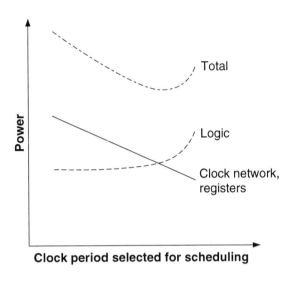

Figure 5.13. Effect of varying the clock period on power consumption

network needs to be charged and discharged fewer times, leading to smaller power consumption in the clock network and registers.

- Larger clock periods may lead to implementations with smaller controllers (due to fewer states and state transitions), and hence a reduction in controller power.

- The increased functional unit chaining that results from larger clock period values serves to inhibit functional unit sharing, which may sometimes lead to larger and more power consuming data paths.

- Larger clock period values may result in higher clock slacks due to granularity effects [48], potentially requiring faster, but more switched-capacitance-hungry, functional units to meet the same performance or inhibiting the application of supply voltage scaling.

Some of the above trade-offs are illustrated graphically in Figure 5.13, which shows the variation of the logic, clock network and register, and total power consumption in a typical RTL implementation as a function of the clock period used during scheduling. For large clock periods, the large glitching power consumption in the logic due to the excessive use of resource chaining dominates the logic power. As the clock period is decreased, glitching power decreases, however the clock distribution network and register power increases. At very low clock periods, the increase in clock and register power outweighs the savings in the logic power consumption. The effects of this trade-off can be explored at the logic level using retiming and pipelining [122, 123]. In order to explore this trade-off during high-level synthesis, accurate power estimation techniques for the clock network, data path, control logic, and registers, such as those presented in Chapter 3, are required.

5.5 SUPPLY VOLTAGE VS. SWITCHED CAPACITANCE TRADE-OFFS

Most of the techniques presented in this chapter deal with the optimization of switched capacitance and supply voltage separately. In general, the optimizations of switched capacitance and supply voltage are strongly correlated, leading to a switched capacitance *vs.* supply voltage trade-off. Some of the factors behind this trade-off are as follows.

- The use of performance optimizing transformations to enable supply voltage scaling may lead to an increase in the number of operations, or the use of more energy consuming operations.

- Faster library components, which are used to implement operations on the critical paths of the DFG in order to enable supply voltage reduction, incur higher switched capacitance.

140   HIGH-LEVEL POWER ANALYSIS AND OPTIMIZATION

- Exploitation of concurrency to reduce the supply voltage inhibits resource sharing, and may lead to larger clock networks (due to larger chip area) which implies an increase in the switched capacitance of the clock network.

The supply voltage *vs.* switched capacitance trade-off is illustrated by the following example.

Example 5.11 *Figure 5.14(a) shows an example DFG which represents the computation of the dot product of two vectors in the Cartesian form. The operations have been mapped to the fastest available functional units. The DFG in Figure 5.14(a) is annotated with the scheduling and resource sharing information, using dashed lines to indicate clock edges or control step boundaries, and dashed ellipses to group operations that share the same functional units. Suppose the*

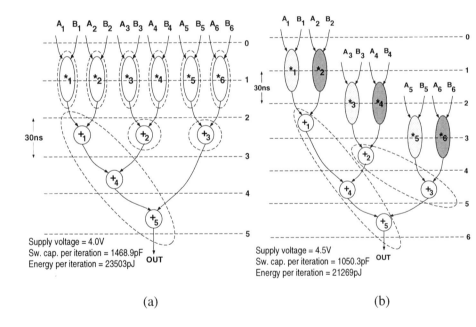

Figure 5.14.   Supply voltage *vs.* switched capacitance trade-off

performance constraint required is that each iteration of the DFG be completed in $200ns$. The schedule shown in Figure 5.14(a) completes in $30 * 5 = 150ns$. Since the schedule completes faster than required by the performance constraint, the supply voltage can be reduced to $4V$ while meeting the performance constraint.

An alternative implementation for the same DFG is shown in Figure 5.14(b). Two-stage pipelined multipliers are used to perform the multiplications and the schedule is increased by one control step. As a result, the number of multipliers required is reduced to two. The multiplication operations in Figure 5.14(b) have been shaded to indicate the multiplier they are assigned to. The schedule shown in Figure 5.14(b) completes in $30 * 6 = 180ns$. Based on the same sample period constraint as before, the supply voltage can be reduced to $4.5V$.

Based on a comparison of the supply voltage, the implementation of Figure 5.14(a) appears to be better since it can be run at a lower supply voltage. However, also considering the effect of switched capacitance changes the scenario. The switched capacitance per DFG iteration for the architectures corresponding to Figures 5.14(a) and 5.14(b) (obtained from complete layouts of both the architectures using a 1.2μ process, a switch-level simulator and pseudo-random input sequences) are $1468.94pf$ and $1050.30pf$, respectively. Using the switched capacitance and supply voltage numbers, the energy consumed per iteration for the architectures of Figures 5.14(a) and 5.14(b) can be computed to be $23503pJ$ and $21269pJ$, respectively. Consequently, the architecture derived from Figure 5.14(b) has a lower power consumption, though it uses a higher supply voltage, since it has lower switched capacitance.

As illustrated by the previous example, it is important to consider the effects of the different high-level synthesis subtasks on both supply voltage and switched capacitance in order to truly minimize power consumption. High-level synthesis optimizations that simultaneously consider the effects on both switched capacitance and supply voltage are presented in [30, 32].

5.6 OPTIMIZING MEMORY POWER CONSUMPTION DURING HIGH-LEVEL SYNTHESIS

The power requirements of many applications, including several multimedia and telecommunication applications, are dominated by the power consumed by storing and retrieving data from memories. The high-level synthesis process maps arrays in the behavioral description to physical on-chip or off-chip memories in the implementation, and synthesizes circuitry, where necessary, to generate addresses to access the memories. Power trade-offs involved during this process include the following.

- Applying loop transformations to reduce the number of memory accesses and required memory capacities (*e.g.* loop reordering, loop merging, nested loop interchange, *etc.*) [8].

- Determining the number, types, and capacities of physical memories to use in the implementation (*e.g.* shared *vs.* distributed memories, single-port *vs.* multi-port memories, *etc.*) [124, 125].

- Defining memory hierarchies [126].

- Mapping of the arrays in the behavioral description to the physical memories, and assigning ports through which they are accessed in the case of multiport memories (*e.g.* row-major mapping *vs.* column-major mapping, array interleaving, *etc.*) [9, 127, 128].

Some of the above trade-offs are illustrated through examples below.

Example 5.12 *This example illustrates how loop transformations can be used to reduce the number of memory accesses required, and also the memory size. Consider the pseudocode shown in Figure 5.15(a) that represents part of a behavioral description that uses three arrays A, B, and C, of size N. Further, let us assume that array B is not used anywhere else in the behavioral description. Since the*

```
...                                                 ...                     ...
...                                                 ...                     ...
...                         ...                     ...                     ...
FOR (i := 1 TO N) DO        ...                     FOR (i := 1 TO N) DO    FOR (i := 1 TO N) DO
  B[i] := f(A[i]);          FOR (i := 1 TO N) DO      B[i] := f(A[i]);        D := g(C[i],D);
END;                          B[i] := f(A[i]);     END;                    END;
FOR (i := 1 TO N) DO          C[i] := g(B[i]);     FOR (i := 1 TO N) DO    FOR (i := 1 to N) DO
  C[i] := g(B[i]);          END;                     D := g(C[i],D);         B[i] := f(A[i]);
END;                        ...                    END;                    END;
...                         ...                    ...                     ...
...                         ...                    ...                     ...
...                         ...                    ...                     ...

         (a)                         (b)                    (c)                     (d)
```

Figure 5.15. Loop transformations for optimizing memory size and number of memory accesses [8]

two loops have similar iteration patterns, they can be merged, as shown in the pseudocode of Figure 5.15(b). As a result of this transformation, the value of each of the entries of array B is used immediately after it is generated. Hence, we do not require any memory storage for array B; a register is used to store the required value and its contents are used immediately. The number of memory accesses is also reduced by $2 * N$, since the generation and use of the entries of array B no longer correspond to memory accesses.

Next, consider the partial behavioral description given by the pseudocode in Figure 5.15(c). Assume that A and B are used in the behavioral description after the loops shown in the figure, and that C is not used later. In this case, we require separate physical memory locations for arrays A, B, and C. The loops can be re-ordered, as shown in the pseudocode of Figure 5.15(d). The benefit of this transformation is that array B can use the same physical memory locations as array C, reducing the size of the physical memory required.

Example 5.13 *This example illustrates the effect of mapping arrays to physical memory locations on the power consumption in the memory address bus circuitry. This optimization is especially important when the memory resides off-chip, incur-*

144   HIGH-LEVEL POWER ANALYSIS AND OPTIMIZATION

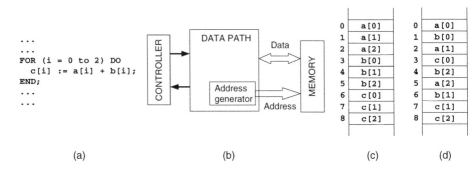

Figure 5.16. Mapping arrays to memories in order to minimize transitions on the address bus [9]

ring power consumption in I/O drivers and large off-chip bus capacitances when it is accessed. Consider the partial behavioral description that is given in Figure 5.16(a) that uses three arrays a, b, and c of size three. The sequence in which the array elements are accessed is $(a[0], b[0], c[0], a[1], b[1], c[1], a[2], b[2], c[2])$. The architectural template of the RTL implementation is shown in Figure 5.16(b). The implementation consists of a controller, a data path that contains address generation circuitry that generates the values to be written onto the memory address bus, and the memory. Consider two candidate mappings of the arrays a, b, and c to physical memory locations. In the case of sequential mapping, which is shown in Figure 5.16(c), the sequence of values appearing on the memory address bus is $(0, 3, 6, 1, 4, 7, 2, 5, 8)$. This results in a total of 19 transitions on the address bus, assuming that no address bus encoding scheme is employed. On the other hand, in the case of the mapping shown in Figure 5.16(d), the memory address bus sequence is $(0, 1, 3, 2, 6, 7, 5, 4, 8)$, which results in a total of 9 transitions. Similar savings can be achieved even when Gray code addressing is employed for the memory address bus [9].

5.7 REDUCING GLITCHING POWER CONSUMPTION DURING HIGH-LEVEL DESIGN

As shown in Chapter 3, glitching accounts for a significant portion of the total power consumption, making it important to consider the effects of glitching power while performing high-level trade-offs. For example, the extent of resource chaining permitted during scheduling affects the glitching power in the data path as mentioned in Section 5.4.2. However, a reasonably accurate analysis of glitching power, using techniques such as those presented in Chapter 3 requires a knowledge of some details about the circuit structure. The structural details of the RTL circuit are available during the later steps of the high-level synthesis flow, *e.g.* during or after resource sharing. However, complete details about the structure, *e.g.* the decomposition of large multiplexers into trees of smaller (*e.g.* 2-to-1) multiplexers may not be available. Transformations that can be applied to such RTL circuits to minimize the power consumption due to the generation and propagation of glitches are presented in [108, 129], including:

- Multiplexer decomposition and multiplexer tree structuring to eliminate the use of glitchy control signals, and minimize glitch propagation from data and control signals.

- Selective delay insertion to minimize glitch propagation.

- Using the clock signal to suppress glitchy transitions

- Architectural delay balancing using buffers and transparent latches.

The following example illustrates how ignoring glitches can be misleading and result in designs that are sub-optimal in terms of their power consumption.

Example 5.14 *Consider the two RTL architectures shown in Figures 5.17(a) and 5.17(b). Both architectures implement the simple function:* if(x < y) then z = c + d else z = a + b. ARCHITECTURE 2 *uses more resources than* ARCHITECTURE 1 *since the former uses two adders as opposed to one adder for the*

146   HIGH-LEVEL POWER ANALYSIS AND OPTIMIZATION

latter. Based on the number of operations performed, a metric that is commonly used to estimate power consumption at the behavior and architecture levels, it seems that ARCHITECTURE 2 *would consume more power than* ARCHITECTURE 1. *This conclusion is supported by power estimation results which do not take glitches into account. However, when accurate power estimation that also considers glitches is performed, it turns out that* ARCHITECTURE 2 *actually consumes 17.7% less power than* ARCHITECTURE 1.

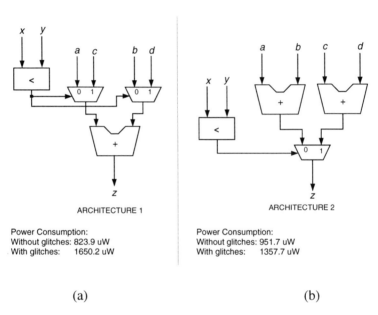

Figure 5.17. Alternative architectures that implement the same function: Effect of glitching

The above observation can be explained as follows. The comparator generates glitches at its output though its inputs are glitch-free. In the case of ARCHITECTURE 1, *these glitches then propagate through the two multiplexers to the inputs of the adder, which causes a significant increase in glitching activity and hence power consumption in the two multiplexers and the adder. In* ARCHITECTURE 2, *though the comparator generates glitches as before, the effect of these glitches is restricted to the single multiplexer.*

The insights obtained from the previous example can be utilized during resource sharing when deciding whether or not to share operations whose execution is dependent on the evaluation of a conditional within the same control step. Sharing such mutually exclusive operations leads to control dependencies, as in the case of ARCHITECTURE 1 of Figure 5.17, potentially leading to glitching power consumption in the shared functional unit.

Multiplexer and control logic transformations for reducing glitching power

As shown in Chapter 3, significant glitching activity can be generated at the control signals. These glitches can propagate through the other parts of the circuit, causing significant power dissipation. The following examples illustrate how to stop glitches on control signals, as close to their source as possible, from propagating further.

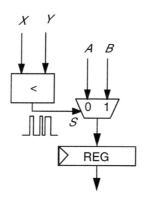

Figure 5.18. Example circuit used to illustrate the effect of data signal correlations on control signal glitches

148   HIGH-LEVEL POWER ANALYSIS AND OPTIMIZATION

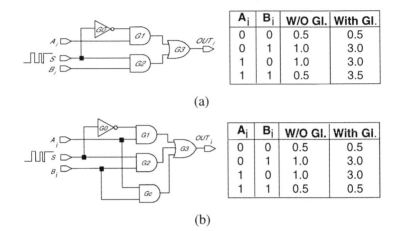

Figure 5.19. (a) Effect of data correlations on select signal glitches, and (b) use of the consensus term to reduce glitch propagation

Example 5.15 *Consider the circuit shown in Figure 5.18. A multiplexer selects between two 8-bit data signals, A and B, depending on whether the expression $X < Y$ evaluates to $True$ or $False$. Its output is written into a register. Suppose that the less-than comparator generates glitches at its output, and that data inputs to the multiplexer are not glitchy and settle to their final value well before the select signal settles. The glitches on the select signal of the multiplexer propagate to its output. In order to study this propagation, consider the gate-level implementation of a bit-slice of the multiplexer that is shown in Figure 5.19(a). The table shown in Figure 5.19(a) reports the glitches at the multiplexer output for all possible values of the data signal bits A_i and B_i. In this table, a rising or falling transition is counted as a half-transition. In the $<0,0>$ case, glitches on select signal S are killed at* AND *gates G1 and G2 due to controlling side inputs that arrive early. When data inputs are $<0,1>$ ($<1,0>$), glitches on S propagate through gates G2 and G3 (G1 and G3). Finally, when data inputs are $<1,1>$, glitches on S propagate through gates G1 and G2. The output of the multiplexer is glitchy as a* result of the interaction of the glitchy signal waveforms at $G1$ and $G2$. The exact manner in which the waveforms interact depends on the propagation and inertial delays of the various wires and gates in the implementation. There are many ways of preventing the propagation of glitches for the $<1,1>$ case. One way is to add an extra gate Gc, as shown in Figure 5.19(b). Gc realizes $A_i.B_i$ which is the consensus of $\bar{S}.A_i$ and $S.B_i$. When data inputs are $<1,1>$, Gc effectively kills any glitches at the other inputs of $G3$ that arrive after the output of Gc settles to a 1, as shown in the table of Figure 5.19(b). Maximum benefits are derived from the addition of the consensus term when the select signal is very glitchy, the data inputs arrive early compared to the select signal, and the probability of the data inputs being $<1,1>$ is high.

Note that with the addition of the consensus term, glitches do not propagate from the select signal to the multiplexer output if the data values are correlated ($<0,0>$ or $<1,1>$). The next example shows how to restructure a multiplexer tree so as to maximize data correlations and hence minimize propagation of glitches from its select signals.

Example 5.16 *Consider the 3-to-1 multiplexer network that is shown in Figure 5.20(a). Functionally, the multiplexer tree can be thought of as an abstract 3-to-1 multiplexer, as shown in Figure 5.20(b). The conditions under which $OUTPUT$, X and $ZERO$ are selected are represented as C_{OUTPUT}, C_X, and C_{ZERO}, respectively. Note that C_{OUTPUT}, C_X, and C_{ZERO} must be mutually exclusive. The cumulative switching activities with and without glitches are shown for various signals in the figure.*

Given the abstract representation of the 3-to-1 multiplexer network there are several possible implementations which enhance correlations of the data inputs to the multiplexers in the tree. For this example, select signal C_{ZERO} is observed to be glitchy, leading to propagation of glitches to the output of the first 2-to-1 multiplexer in Figure 5.20(a). Note that data signals $OUTPUT$ and $ZERO$ are highly correlated at the bit level. In order to minimize the propagation of glitches

150 HIGH-LEVEL POWER ANALYSIS AND OPTIMIZATION

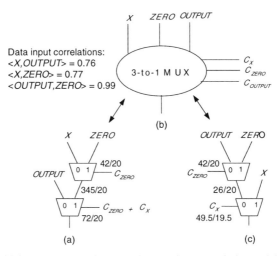

Figure 5.20. Multiplexer restructuring to enhance data correlations: (a) initial multiplexer network, (b) abstract 3-to-1 multiplexer, and (c) restructured network

on C_{ZERO} through the multiplexer tree, the multiplexer tree is transformed to the implementation shown in Figure 5.20(c), such that the highly correlated data signals $OUTPUT$ and $ZERO$ become inputs to the first 2-to-1 multiplexer. This significantly lowers the switching activity at the output of the first 2-to-1 multiplexer to $26/20$ from $345/20$ originally.

The previous examples addressed minimizing glitch propagation from the control logic. The following examples illustrate techniques to minimize the glitch propagation from data signals, such as outputs of functional units and multiplexers.

Example 5.17 *Consider the example circuit shown in Figure 5.21(a). A 2-to-1 multiplexer selects between the outputs of two adders, and the multiplexer's output is fed to another adder. This is a situation that occurs commonly in RTL designs that employ data chaining. The adders feeding the multiplexer generate glitches even when their inputs are glitch-free, leading to glitch propagation through the*

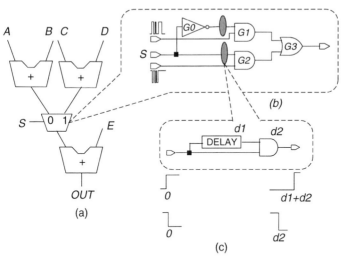

Figure 5.21. (a) Example circuit, (b) multiplexer bit-slice with selective delays inserted, and (c) implementation of a rising delay block

multiplexer and then through the third adder, causing significant power dissipation. A technique called *selective delay insertion* can be used to cut down the propagation of glitches through the circuit, as explained next.

Consider the gate-level implementation of a bit-slice of the multiplexer as shown in Figure 5.21(b). Both the data inputs to the multiplexer are glitchy. Consider a pair of consecutive clock cycles q_1 and q_2 such that the select signal to the multiplexer makes a $1 \rightarrow 0$ (falling) transition from q_1 to q_2. If the falling transition at S is early arriving, there will be an early rising transition at the output of gate $G0$ that implements $\bar{S}$. Consequently, the side input of $G1$ will become non-controlling early, allowing the data input glitches to propagate through $G1$. This propagation can be minimized by ensuring that the side input to $G1$ remains controlling as long as possible, which can be achieved by adding a "selective rising transition delay" to the output of $G0$ ($\bar{S}$). Similarly, to minimize glitch propagation through gate $G2$ when there is an early rising transition at S, it is *desirable to delay the rising transition on the fanout branch of S that feeds* AND *gate $G2$. The selective rising delay blocks are represented by the shaded ellipses shown in Figure 5.21(b). A possible implementation of a rising delay block, that uses one* AND *gate and a delay element, is shown in Figure 5.21(c). The delay element is constructed using either a series of buffers or inverters added to the input. The implementation uses the fact that a falling transition at any one input of an* AND *gate is sufficient to force the output to 0, while, on the other hand, the latest arriving rising transition among all the inputs will trigger a rising transition at the output. Under a simplified delay model of d_1 ns for the delay block and d_2 ns for the* AND *gate, it can be seen that a rising transition at the input is delayed by $(d_1 + d_2)$ ns, while a falling transition is delayed by only d_2 ns. Since d_1 is typically large compared to d_2, the slight increase in propagation of glitches due to the additional delay of d_2 ns imparted to the falling transition is far outweighed by the savings obtained for the case of a rising transition. A selective falling delay block is similar to the circuit shown in Figure 5.21(c), except that the* AND *gate is replaced by an* OR *gate. Note that for an entire m-bit multiplexer, it suffices to have selective rising delays at the select signal S and its complement $\bar{S}$. To allow this low-cost solution, an m-bit selector is used instead of a multiplexer (a selector implements the function $S_1.A + S_2.B$). The two select signals $(S_1$ and $S_2)$ are generated explicitly outside the selector as S and $\bar{S}$. Applying the above technique to the example circuit shown in Figure 5.21(a) results in a 15.4% decrease in overall power consumption, estimated by a gate-level power simulator using a technology-mapped netlist. However, in general, it is necessary to consider the power consumed by the selective delay block itself, the probability of the relevant transition occurring at the delay block's input, and the impact of inserting the delay block on the circuit delay.*

5.8 CONCLUSIONS

This chapter illustrated the effect of various high-level synthesis subtasks on power consumption and presented techniques that can be used to minimize power consumption when performing the subtasks, including behavioral transformations, module selection, clock period selection, resource sharing to exploit data correlations and regularity, scheduling, memory optimizations, and RTL circuit transformations. These techniques have been demonstrated by several researchers to yield power savings that are much larger than those attainable using lower-level optimizations. However, most of the techniques are best suited to data-flow intensive applications, where power consumption is dominated either by the functional units in the data path or by the memory operations. An important direction of future work is the development of high-level synthesis for low power techniques that are applicable to control-flow intensive designs, and designs that contain a significant mix of data flow and control flow. Another challenge arises due to the fact that the effects of the various high-level synthesis tasks on power are interdependent. In order to realize maximal power savings, it is important to come up with a comprehensive high-level framework that performs the various optimizations in a coherent and integrated manner.

6 CONCLUSIONS AND FUTURE WORK

Power consumption has been elevated to become one of the most important metrics in the design of electronic circuits and systems, for a variety of reasons that were discussed in Chapter 1. While significant process technology improvements do result in large power reductions, they are by no means sufficient to address the need for low power design. As a result, power analysis and optimization tools that operate at various levels of the design hierarchy are needed. Most research and commercial development work to date has focussed on developing power estimation and minimization tools at the lower (transistor and logic) levels of the design hierarchy that offer push-button power savings during the later steps in a design flow. This book presented power analysis and optimization techniques for designs represented at the higher (algorithm and architecture) levels of the design hierarchy. The techniques and results presented in this book demonstrate that various high-level and RTL transformations do have a significant impact on power consumption. Further, it has been shown that efficient exploration of the design space at the higher levels leads to large power savings, far beyond those that can be obtained through logic and circuit optimizations. In light of the increasing importance of power consumption as a design metric, the availability and use of high-level power analysis and optimization tools will enable the designers to meet very stringent power constraints, and can lead to fewer and faster design iterations. Hence, the incorporation of high-level power analysis and optimization methodologies and techniques, such as those presented in this book, into commercial EDA tools is an immediate requirement. Some of the initial work in this direction has been incorporated into tools such as Watt Watcher from Sente, Inc. that performs architectural power estimation.

There are several related issues of interest that can be explored in the future. Each of the chapters has outlined some of the specific improvements that could be made to the techniques presented in this book. Besides such opportunities, there will be a need in the near future for design tools that support levels of abstraction that are even higher than those dealt with in this book. Advances in VLSI process technology has led to the possibility of integration of entire systems, that hitherto consisted of several chips on a board, onto a single chip. Such systems typically consist of "embedded" programmable microprocessors running software that interact with hardware. It is estimated that the market for system-on-a-chip ASICs will grow from $1.1 billion or 10% of the entire ASIC market in 1996 to $14 billion or 60% of the ASIC market in 2000 (Source: DataQuest, Inc.). There is currently a large void of design tools and techniques, in general, and power analysis and optimization tools in particular, to aid system-level design tasks and decisions, such as partitioning of functionality into hardware and software, hardware/software communication strategies, the choice of operating system and multiprocessing strategies for the software, *etc.* The study of previously developed lower-level power analysis and optimization techniques provided significant insights that helped develop the techniques presented in this book. Similarly, it is our hope that the insights and techniques developed in this book will aid in the development of power-sensitive system-level design tools and methodologies.

REFERENCES

[1] P. Landman and J. M. Rabaey, "Architectural power analysis: The dual bit type method," *IEEE Trans. VLSI Systems*, vol. 3, pp. 173–187, June 1995.

[2] L. Benini, A. Bogliolo, M. Favalli, and G. De Micheli, "Regression models for behavioral power estimation," in *Proc. Int. Wkshp. Power & Timing Modeling, Optimization, and Simulation*, 1996.

[3] A. P. Chandrakasan, M. Potkonjak, R. Mehra, J. Rabaey, and R. Brodersen, "Optimizing power using transformations," *IEEE Trans. Computer-Aided Design*, vol. 14, pp. 12–31, Jan. 1995.

[4] N. Sankarayya, K. Roy, and D. Bhattacharya, "Algorithms for low power FIR filter realization using differential coefficients," in *Proc. Int. Conf. VLSI Design*, pp. 174–178, Jan. 1997.

[5] A. Chatterjee and R. K. Roy, "Synthesis of low power DSP circuits using activity metrics," in *Proc. Int. Conf. VLSI Design*, pp. 265–270, Jan. 1994.

[6] A. Raghunathan and N. K. Jha, "Behavioral synthesis for low power," in *Proc. Int. Conf. Computer Design*, pp. 318–322, Oct. 1994.

[7] R. Mehra and J. Rabaey, "Exploiting regularity for low-power design," in *Proc. Int. Conf. Computer-Aided Design*, pp. 166–172, Nov. 1996.

[8] S. Wuytack, F. Catthoor, F. Franssen, L. Nachtergaele, and H. De Man, "Global communication and memory optimizing transformations for low power systems," in *Proc. Int. Wkshp. Low Power Design*, pp. 203–208, Apr. 1994.

[9] P. R. Panda and N. D. Dutt, "Reducing address bus transitions for low power memory mapping," in *Proc. European Design & Test Conf.*, pp. 63–67, Mar. 1996.

[10] J. Eager, "Advances in rechargeable batteries spark product innovation," in *Proc. Silicon Valley Computer Conf.*, pp. 243–253, Aug. 1992.

[11] D. Maliniak, "Better batteries for low-power jobs," *Electronic Design*, vol. 40, p. 18, July 1992.

[12] J. Rabaey and M. Pedram (Editors), *Low Power Design Methodologies*. Kluwer Academic Publishers, Norwell, MA, 1996.

[13] C. Small, "Shrinking devices put the squeeze on system packaging," *Electronic Design News*, vol. 39, pp. 41–46, Feb. 1994.

[14] H. B. Bakoglu, *Circuits, Interconnections, and Packaging for VLSI*. Addison-Wesley, Menlo Park, CA, 1990.

[15] P. Yang and J. H. Chern, "Design for reliability: The major challenge for VLSI," *Proc. IEEE*, vol. 81, pp. 730–744, May 1993.

[16] J.-F. Tuan and T. K. Young, "Reliability issues in power and ground on submicron circuit," in *Proc. WESCON*, pp. 129–133, Nov. 1995.

[17] B. Nadel, "The green machine," *PC Magazine*, vol. 12, May 1993.

REFERENCES

[18] D. Yuen, *Intel's SL Architecture - Designing Portable Applications*. McGraw-Hill, New York, NY, 1993.

[19] D. L. Perry, *VHDL*. McGraw-Hill, New York, NY, 1991.

[20] A. R. Chandrakasan and R. W. Brodersen, *Low Power Digital CMOS Design*. Kluwer Academic Publishers, Norwell, MA, 1995.

[21] J. Frenkil, "Tools and methodologies for low power design," in *Proc. Design Automation Conf.*, pp. 76–81, June 1997.

[22] W. Nebel, J. Sproch, and S. Malik, "Tutorial: Power analysis and optimization: spanning the levels of abstraction," in *Proc. Int. Symp. Low Power Electronics & Design*, Aug. 1997.

[23] A. Bellaouar and M. I. Elmasry, *Low-Power Digital VLSI Design - Circuits and Systems*. Kluwer Academic Publishers, Norwell, MA, 1995.

[24] N. H. E. Weste and K. Eshraghian, *Principles of CMOS VLSI design, 2nd edition*. Addison-Wesley, Menlo Park, CA, 1994.

[25] M. Horowitz, T. Indermaur, and R. Gonzalez, "Low-power digital design," in *Proc. Symp. Low Power Electronics*, pp. 8–11, Oct. 1994.

[26] A. Stratakos, R. W. Brodersen, and S. R. Sanders, "High-efficiency low-voltage DC-DC conversion for portable applications," in *Proc. Int. Wkshp. Low Power Design*, pp. 105–110, Apr. 1994.

[27] K. Seta, H. Hara, T. Kuroda, M. Kakumu, and T. Sakurai, "50% active power reduction saving without speed degradation using standby power reduction (SPR) circuit," in *Proc. Int. Solid-State Circuits Conf.*, pp. 318–319, Feb. 1996.

[28] L. Goodby, A. Orailoglu, and P. M. Chau, "Microarchitectural synthesis of performance-constrained, low-power VLSI designs," in *Proc. Int. Conf. Computer Design*, pp. 323–326, Oct. 1994.

[29] R. S. Martin and J. P. Knight, "Power Profiler: Optimizing ASICs power consumption at the behavioral level," in *Proc. Design Automation Conf.*, pp. 42–47, June 1995.

[30] A. Raghunathan and N. K. Jha, "An iterative improvement algorithm for low power data path synthesis," in *Proc. Int. Conf. Computer-Aided Design*, pp. 597–602, Nov. 1995.

[31] A. P. Chandrakasan, S. Sheng, and R. W. Brodersen, "Low-power CMOS digital design," *IEEE J. Solid-State Circuits*, vol. 27, pp. 473–484, Apr. 1992.

[32] R. Mehra and J. Rabaey, "Behavioral level power estimation and exploration," in *Proc. Int. Wkshp. Low Power Design*, pp. 197–202, Apr. 1994.

[33] S. Raje and M. Sarrafzadeh, "Variable voltage scheduling," in *Proc. Int. Symp. Low Power Design*, pp. 9–14, Apr. 1995.

[34] J. M. Chang and M. Pedram, "Energy minimization using multiple supply voltages," in *Proc. Int. Symp. Low Power Electronics & Design*, pp. 157–162, Aug. 1996.

[35] M. Johnson and K. Roy, "Optimal selection of supply voltages and level conversions during data path scheduling under resource constraints," in *Proc. Int. Conf. Computer Design*, pp. 72–77, Oct. 1996.

[36] T. Blalock and J. Jaeger, "A high-speed clamped bitline current-mode sense amplifier," *IEEE J. Solid-State Circuits*, vol. 26, pp. 542–548, Apr. 1991.

[37] D. Somasekhar and K. Roy, "Differential current switch logic: A low power DCVS logic family," *IEEE J. Solid-State Circuits*, vol. 31, pp. 981–991, July 1996.

[38] M. Pedram and H. Vaishnav, "Power optimization in VLSI layout: A survey," *J. VLSI Signal Processing*, 1996.

[39] S. Devadas and S. Malik, "A survey of optimization techniques targeting low power VLSI circuits," in *Proc. Design Automation Conf.*, pp. 242–247, June 1995.

[40] J. Monteiro and S. Devadas, *Computer-Aided Design Techniques for Low Power Sequential Logic Circuits*. Kluwer Academic Publishers, Norwell, MA, 1996.

[41] M. Pedram, "Power minimization in IC design: Principles and applications," *ACM Trans. Design Automation Electronic Systems*, vol. 1, pp. 3–56, Jan. 1996.

[42] V. Tiwari, S. Malik, and A. Wolfe, "Compilation techniques for low energy: An overview," in *Proc. Symp. Low Power Electronics*, pp. 38–39, Oct. 1994.

[43] V. Tiwari, S. Malik, A. Wolfe, and T. C. Lee, "Instruction level power analysis and optimization of software," in *Proc. Int. Conf. VLSI Design*, pp. 326–328, Jan. 1996.

[44] D. D. Gajski, N. D. Dutt, A. C.-H. Wu, and S. Y.-L. Lin, *High-level Synthesis: Introduction to Chip and System Design*. Kluwer Academic Publishers, Norwell, MA, 1992.

[45] P. G. Paulin, J. P. Knight, and E. F. Girczyc, "HAL: A multi-paradigm approach to automatic data path synthesis," in *Proc. Design Automation Conf.*, pp. 263–270, June 1986.

[46] D. E. Thomas and G. Leive, "Automatic technology relative logic synthesis and module selection," *IEEE Trans. Computer-Aided Design*, vol. 2, pp. 94–105, Apr. 1983.

[47] G. De Micheli, *Synthesis and Optimization of Digital Circuits*. McGraw-Hill, New York, NY, 1994.

[48] S. Narayanan and D. D. Gajski, "System clock estimation based on clock slack minimization," in *Proc. European Design Automation Conf.*, pp. 66–71, Feb. 1992.

[49] S. Chaudhuri, S. A. Blythe, and R. A. Walker, "An exact solution methodology for scheduling in a 3D design space," in *Proc. Int. Symp. System Level Synthesis*, pp. 78–83, Sept. 1995.

[50] M. Corazao, M. Khalaf, L. Guerra, M. Potkonjak, and J. Rabaey, "Instruction set mapping for performance optimization," in *Proc. Int. Conf. Computer-Aided Design*, pp. 518–521, Oct. 1993.

[51] B. M. Pangrle and D. D. Gajski, "Design tools for intelligent silicon compilation," *IEEE Trans. Computer-Aided Design*, vol. 6, pp. 1098–1112, June 1987.

[52] P. G. Paulin and J. P. Knight, "Force-directed scheduling for the behavioral synthesis of ASIC's," *IEEE Trans. Computer-Aided Design*, vol. 8, pp. 661–679, June 1989.

[53] S. Devadas and A. R. Newton, "Algorithms for hardware allocation in data path synthesis," *IEEE Trans. Computer-Aided Design*, vol. 8, pp. 768–781, July 1989.

[54] I. C. Park and C. M. Kyung, "FAMOS: An efficient scheduling algorithm for high-level synthesis," *IEEE Trans. Computer-Aided Design*, vol. 12, pp. 1437–1448, Oct. 1993.

[55] J. H. Lee, Y. C. Hsu, and Y. L. Lin, "A new integer linear programming formulation for the scheduling problem in data path synthesis," in *Proc. Int. Conf. Computer-Aided Design*, pp. 20–23, Nov. 1989.

[56] C. H. Gebotys and M. I. Elmasry, "A global optimization approach for architectural synthesis," in *Proc. Int. Conf. Computer-Aided Design*, pp. 258–261, Nov. 1990.

REFERENCES   165

[57] C. T. Hwang et al., "PLS: Scheduler for pipeline synthesis," *IEEE Trans. Computer-Aided Design*, vol. 12, pp. 1279–1286, Sept. 1993.

[58] R. Camposano, "Path-based scheduling for synthesis," *IEEE Trans. Computer-Aided Design*, vol. 10, pp. 85–93, Jan. 1991.

[59] S. Bhattacharya, S. Dey, and F. Brglez, "Performance analysis and optimization of schedules for conditional and loop-intensive specifications," in *Proc. Design Automation Conf.*, pp. 491–496, June 1994.

[60] G. Lakshminarayana, K. S. Khouri, and N. K. Jha, "Wavesched: A novel scheduling technique for control-flow intensive behavioral descriptions," in *Proc. Int. Conf. Computer-Aided Design*, Nov. 1997.

[61] S. Bhattacharya, F. Brglez, and S. Dey, "Transformations and resynthesis for testability of RT-level control-data path specifications," *IEEE Trans. VLSI Systems*, vol. 1, pp. 304–318, Sept. 1993.

[62] E. A. Rundensteiner, D. D. Gajski, and L. Bic, "Component synthesis from functional descriptions," *IEEE Trans. Computer-Aided Design*, vol. 12, pp. 1287–1299, Sept. 1993.

[63] F. Kurdahi and A. C. Parker, "Techniques for area estimation of VLSI layouts," *IEEE Trans. Computer-Aided Design*, vol. 8, pp. 81–92, Jan. 1989.

[64] P. K. Jha and N. D. Dutt, "Rapid estimation for parameterized components in high-level synthesis," *IEEE Trans. VLSI Systems*, vol. 1, pp. 296–303, Sept. 1993.

[65] C. Ramachandran and F. J. Kurdahi, "Incorporating the controller effects during register-transfer level synthesis," in *Proc. European Design & Test Conf.*, pp. 308–313, Mar. 1994.

[66] S. Bhattacharya, S. Dey, and F. Brglez, "Provably correct high-level timing analysis without path sensitization," in *Proc. Int. Conf. Computer-Aided Design*, pp. 736–742, Nov. 1994.

[67] A. Raghunathan, S. Dey, and N. K. Jha, "Register-transfer level estimation techniques for switching activity and power consumption," in *Proc. Int. Conf. Computer-Aided Design*, pp. 158–165, Nov. 1996.

[68] K. D. Müller-Glaser, K. Kirsch, and K. Neusinger, "Estimating essential design characteristics to support project planning for ASIC design management," in *Proc. Int. Conf. Computer-Aided Design*, pp. 148–151, Nov. 1991.

[69] D. Liu and C. Svensson, "Power consumption estimation in CMOS VLSI chips," *IEEE J. Solid-State Circuits*, vol. 29, pp. 663–670, June 1994.

[70] D. Marculescu, R. Marculescu, and M. Pedram, "Information theoretic measures for energy consumption at the register-transfer level," in *Proc. Int. Symp. Low Power Design*, pp. 81–86, Apr. 1995.

[71] F. N. Najm, "Towards a high-level power estimation capability," in *Proc. Int. Symp. Low Power Design*, pp. 87–92, Apr. 1995.

[72] N. Pippinger, "Information theory and the complexity of Boolean functions," *Mathematical Systems Theory*, vol. 10, pp. 129–167, 1977.

[73] K.-T. Cheng and V. D. Agrawal, "An entropy measure for the complexity of multi-output Boolean functions," in *Proc. Design Automation Conf.*, pp. 302–305, June 1990.

[74] M. Nemani and F. N. Najm, "High-level power estimation and the area complexity of Boolean functions," in *Proc. Int. Symp. Low Power Electronics & Design*, pp. 329–334, Aug. 1996.

[75] S. R. Powell and P. M. Chau, "Estimating power dissipation of VLSI signal processing chips: The PFA technique," in *Proc. VLSI Signal Processing IV*, pp. 250–259, Sept. 1990.

[76] P. E. Landman and J. M. Rabaey, "Black-box capacitance models for architectural power analysis," in *Proc. Int. Wkshp. Low Power Design*, pp. 165–170, Apr. 1994.

[77] P. Landman and J. M. Rabaey, "Activity-sensitive architectural power analysis for the control path," in *Proc. Int. Symp. Low Power Design*, pp. 93–98, Apr. 1995.

[78] T. Sato, Y. Ootaguro, M. Nagamatsu, and H. Tago, "Evaluation of architecture-level power estimation for CMOS RISC processors," in *Proc. Symp. Low Power Electronics*, pp. 44–45, Oct. 1995.

[79] S. Gupta and F. N. Najm, "Power macromodeling for high level power estimation," in *Proc. Design Automation Conf.*, pp. 365–370, June 1997.

[80] *CMOS6 Library Manual*. NEC Electronics, Inc., Dec. 1992.

[81] *CSIM Version 5 Users Manual*. Systems LSI Division, NEC Corp., 1993.

[82] G. Casella and R. L. Berger, *Statistical Inference*. Duxbury Press, Belmont, CA, 1990.

[83] A. Raghunathan and N. K. Jha, "An ILP formulation for low power based on minimizing switched capacitance during datapath allocation," in *Proc. Int. Symp. Circuits & Systems*, pp. 1069–1073, May 1995.

[84] C.-T. Hsieh and Q. Wu and C.-S. Ding and M. Pedram, "Statistical sampling and regression analysis for RT-level power evaluation," in *Proc. Int. Conf. Computer-Aided Design*, pp. 583–588, Nov. 1996.

[85] H. Mehta and R. M. Owens and M. J. Irwin, "Energy characterization based on clustering," in *Proc. Design Automation Conf.*, pp. 702–707, June 1996.

[86] R. Burch, F. N. Najm, P. Yang, and T. Trick, "A Monte Carlo approach for power estimation," *IEEE Trans. VLSI Systems*, vol. 1, pp. 63–71, Mar. 1993.

[87] D. Marculescu, R. Marculescu, and M. Pedram, "Sequence compaction for probabilistic analysis of finite-state machines," in *Proc. Design Automation Conf.*, pp. 12–15, June 1997.

[88] R. Marculescu, D. Marculescu, and M. Pedram, "Hierarchical sequence compaction for power estimation," in *Proc. Design Automation Conf.*, pp. 570–575, June 1997.

[89] D. I. Cheng and K.-T. Cheng and D. C. Wang and M. Marek-Sadowska, "A new hybrid methodology for power estimation," in *Proc. Design Automation Conf.*, pp. 439–444, June 1996.

[90] P. Landman and J. M. Rabaey, "Activity-sensitive architectural power analysis," *IEEE Trans. Computer-Aided Design*, vol. 15, pp. 571–587, June 1996.

[91] A. Kuehlmann and R. Bergamaschi, "Timing analysis in high-level synthesis," in *Proc. Int. Conf. Computer-Aided Design*, pp. 349–354, Nov. 1992.

[92] L. Benini, P. Siegel, and G. De Micheli, "Saving power by synthesizing gated clocks for sequential circuits," *IEEE Design & Test of Computers*, pp. 32–41, Winter 1994.

[93] L. Benini and G. De Micheli, "Automatic synthesis of gated-clock sequential circuits," *IEEE Trans. Computer-Aided Design*, vol. 15, pp. 630–643, June 1996.

[94] A. Raghunathan, S. Dey, and N. K. Jha, "Register-transfer level power optimization techniques with emphasis on glitch analysis and optimization,"

Tech. Rep. 95-C049-4-5016-3, NEC C&C Research Labs, Princeton, NJ, Oct. 1995.

[95] G. Tellez, A. Farrahi, and M. Sarrafzadeh, "Activity driven clock design for low power circuits," in *Proc. Int. Conf. Computer-Aided Design*, pp. 62–65, Nov. 1995.

[96] C. Papachristou, M. Spining, and M. Nourani, "An effective power management scheme for RTL design based on multiple clocks," in *Proc. Design Automation Conf.*, pp. 337–342, June 1996.

[97] M. Aldina, J. Monteiro, S. Devadas, A. Ghosh, and M. Papaefthymiou, "Precomputation-based sequential logic optimization for low power," *IEEE Trans. VLSI Systems*, vol. 2, pp. 426–436, Dec. 1994.

[98] J. Monteiro, P. Ashar, and S. Devadas, "Scheduling techniques to enable power management," in *Proc. Design Automation Conf.*, pp. 349–352, June 1996.

[99] A. Correale Jr., "Overview of the power minimization techniques employed in the IBM PowerPC 4xx embedded processors," in *Proc. Int. Symp. Low Power Design*, pp. 75–80, Apr. 1995.

[100] V. Tiwari, S. Malik, and P. Ashar, "Guarded evaluation: Pushing power management to logic synthesis/design," in *Proc. Int. Symp. Low Power Design*, pp. 221–226, Apr. 1995.

[101] J. Monteiro, J. Rinderknecht, S. Devadas, and A. Ghosh, "Optimization of combinational and sequential logic circuits for low power using precomputation," in *Proc. Chapel Hill Conf. Advanced Research VLSI*, pp. 430–444, Mar. 1995.

[102] E. Musoll and J. Cortadella, "High-level synthesis techniques for reducing the activity of functional units," in *Proc. Int. Symp. Low Power Design*, pp. 99–104, Apr. 1995.

[103] C. Lee, "Representation of switching circuits by binary decision diagrams," *Bell Systems Tech. J.*, vol. 38, pp. 985–999, July 1959.

[104] S. B. Akers, "Binary decision diagrams," *IEEE Trans. Computers*, vol. C-27, pp. 509–516, June 1978.

[105] M. Abramovici, M. A. Breuer, and A. D. Friedman, *Digital Systems Testing and Testable Design*. New York, NY: Computer Science Press, 1990.

[106] G. Lakshminarayana, A. Raghunathan, N. K. Jha, and S. Dey, "A power management methodology for high-level synthesis," in *Proc. Int. Conf. VLSI Design*, Jan. 1998.

[107] A. Raghunathan, S. Dey, and N. K. Jha, "Power management techniques for control-flow intensive designs," in *Proc. Design Automation Conf.*, pp. 429–434, June 1997.

[108] A. Raghunathan, S. Dey, and N. K. Jha, "Glitch analysis and reduction in register-transfer-level power optimization," in *Proc. Design Automation Conf.*, pp. 331–336, June 1996.

[109] H. Trickey, "Flamel: A high-level hardware compiler," *IEEE Trans. Computer-Aided Design*, vol. 6, pp. 259–269, Mar. 1987.

[110] R. A. Walker and D. E. Thomas, "Behavioral transformations for algorithmic level IC design," *IEEE Trans. Computer-Aided Design*, vol. 8, pp. 1115–1127, Oct. 1989.

[111] B. S. Haroun and M. I. Elmasry, "Architectural synthesis for DSP silicon compilers," *IEEE Trans. Computer-Aided Design*, vol. 8, pp. 431–447, Apr. 1989.

[112] J. Rabaey, C. Chu, P. Hoang, and M. Potkonjak, "Fast prototyping of data path intensive architectures," *IEEE Design & Test of Computers*, vol. 8, pp. 40–51, Feb. 1991.

REFERENCES   171

[113] L. Claesen, F. Catthoor, D. Lanneer, G. Goossens, S. Note, J. V. Meerbergen, and H. De Man, "Automatic synthesis of signal processing benchmark using the CATHEDRAL silicon compilers," in *Proc. Custom Integrated Circuits Conf.*, pp. 14.17.1 – 14.7.4, May 1988.

[114] M. Mehendale, S. D. Sherlekar, and G. Venkatesh, "Coefficient optimizations for low power realization of FIR filters," in *Proc. VLSI Signal Processing VIII*, pp. 352–361, Sept. 1995.

[115] T. Sakurai, H. Kawaguchi, and T. Kuroda, "Low-power CMOS design through V_{TH} control and low-swing circuits," in *Proc. Int. Symp. Low Power Electronics & Design*, pp. 1–6, Aug. 1997.

[116] T. Callaway and E. Swartzlander, "Optimizing arithmetic elements for signal processing," in *Proc. VLSI Signal Processing V*, pp. 91–100, Sept. 1992.

[117] C. Nagendra, R. M. Owens, and M. J. Irwin, "Power-delay characteristics of CMOS adders," *IEEE Trans. VLSI Systems*, vol. 2, pp. 377–381, Sept. 1994.

[118] K. Usami and M. Horowitz, "Clustered voltage scaling technique for low-power design," in *Proc. Int. Symp. Low Power Design*, pp. 3–8, Apr. 1995.

[119] A. Dasgupta and R. Karri, "Simultaneous scheduling and binding for power minimization during microarchitecture synthesis," in *Proc. Int. Symp. Low Power Design*, pp. 69–74, Apr. 1995.

[120] J. M. Chang and M. Pedram, "Register allocation and binding for low power," in *Proc. Design Automation Conf.*, pp. 29–35, June 1995.

[121] J. M. Chang and M. Pedram, "Module assignment for low power," in *Proc. European Design Automation Conf.*, pp. 376–381, Sept. 1996.

[122] J. Monteiro, S. Devadas, and A. Ghosh, "Retiming sequential circuits for low power," in *Proc. Int. Conf. Computer-Aided Design*, pp. 398–402, Nov. 1993.

[123] J. Leijten, J. van Meerbergen, and J. Jess, "Analysis and reduction of glitches in synchronous networks," in *Proc. European Design & Test Conf.*, pp. 398–403, Mar. 1995.

[124] D. Lidsky and J. Rabaey, "Low-power design of memory intensive functions," in *Proc. Symp. Low Power Electronics*, pp. 16–17, Oct. 1994.

[125] D. Lidsky and J. Rabaey, "Low-power design of memory intensive functions case study: Vector quantization," in *Proc. VLSI Signal Processing VII*, pp. 378–387, Sept. 1994.

[126] S. Wuytack, F. Catthoor, L. Nachtergaele, and H. De Man, "Power exploration for data dominated video applications," in *Proc. Int. Symp. Low Power Electronics & Design*, pp. 359–364, Aug. 1996.

[127] P. R. Panda and N. D. Dutt, "Low power mapping of behavioral arrays to multiple memories," in *Proc. Int. Symp. Low Power Electronics & Design*, pp. 289–292, Aug. 1996.

[128] P. R. Panda and N. D. Dutt, "Behavioral array mapping into multiport memories targeting low power," in *Proc. Int. Conf. VLSI Design*, pp. 268–272, Jan. 1997.

[129] C. Lemonds and S. S. Mahant-Shetti, "A low power 16 by 16 multiplier using transition reduction circuitry," in *Proc. Int. Wkshp. Low Power Design*, pp. 139–142, Apr. 1994.

INDEX

Activation function, 84–85
Activity graph, 109–113
Activity-based control, 68
Adaptive macromodeling, 65
Algebraic transformations, 116
Analytical power models, 37–38
ANOVA, 54–55
Application domains, 34
Automatic test pattern generation, 99
Average cube complexity, 41

Barcode pre-processor, 8–9, 26, 29, 32
Battery-driven, 3, 22
Behavioral transformations, 115–116, 119, 123
Binary decision diagram, 99
Bit-level modeling, 57, 59
Bit-line, 38
Boolean difference, 76

Capacitive switching power, 18–20
Cell selection, 24
Chaining, 135, 137–139, 145, 150
Characterization, 42–43, 45, 53, 69
Chip Estimation System, 38
Clock delay, 83
Clock period selection, 29, 115, 137, 153
Clock power, 39–40

Clock skew, 83–84
Clock tree construction, 24, 89–90
Common subexpression elimination, 116, 120
Complete input disabling architecture, 94
Complex number multiplication, 120
Component failure rate, 3
Constant propagation, 116, 118, 120
Constrained register sharing, 102, 106
Control expressions, 51–52, 70, 72
Control flow graph, 27
Control-data flow graph, 29
Control-flow intensive, 26–27, 34–35, 59, 108, 153
Controller-based power management, 107–108, 114
Controlling logic, 72
Cycle-by-cycle behavior, 6, 9, 30–31, 135

Data flow graph, 27
Data-flow intensive, 26, 34–35, 117, 153
Dead code elimination, 116
Design abstraction, 1, 5, 8
Differential cascode voltage switch, 129
Dual bit type, 46
Dynamic variable rebinding, 106

E-template, 133–134
Electromigration, 4
Electronic design automation, 10
Elliptic Wave Filter, 26
Energy clustering, 62
Energy cubes, 62
Energy Star, 5
Entropy, 41–42
Environmental concerns, 5
ESP, 46

Falling delay block, 152
Finite impulse response filter, 121
Firm macroblocks, 43
Functional RTL, 9, 31

Gated clocks, 81–83, 90
Glitching activity, 35, 43, 50–55, 57–58, 67–68, 70, 72–77, 108, 114, 119, 146–147
Greatest common divisor, 51
Ground bounce, 4
Guarded evaluation, 98–99

H-tree, 39
Hamming distance, 131
Handheld electronics, 2
Hard macroblocks, 43
Hardware C, 26
Hot-carrier effects, 4

Infinite impulse response filter, 117
Information transmission coefficient, 42
Information-theoretic approaches, 40
Intermediate format, 27

Layout synthesis, 8, 33
Leakage power, 20–21
Logic synthesis, 5–6, 29, 33, 52, 62, 70, 103
Loop transformations, 116, 142

Mealy FSM, 85
Memory access, 24, 142–143

Memory power, 38, 142
Model decomposition, 55
Module selection, 29, 115, 126, 128–129, 136–137, 153
Monotonicity, 14
Moore FSM, 84–85
Multicycling, 135
Multiple clocks, 24, 81–82, 91
Multiple supply voltages, 23, 128
Multiplexer decomposition, 145

Nickel-Cadmium, 3
Nickel-Metal Hydride, 3

Observability don't care, 98
Operand isolation, 81, 97–98, 100, 102, 107–108

Packaging and cooling costs, 4
Parallel implementations, 123
Partial delay information, 74–75
Peak power, 136–137
Peripheral capacitance model, 59–60, 62–63, 65
Piecewise linear models, 53, 55
Pipelining, 23, 116, 119, 139
Portable systems, 2
Power factor approximation, 45
Power macromodeling, 37, 45, 50, 59
Pre-computation, 81, 93–95
Predictor functions, 93–95

Re-convergent fanout, 132
Regularity, 24, 133–136, 153
Relative accuracy, 14, 50
Reliability issues, 4
Resistive (I-R) voltage drops, 4
Resource allocation, 31
Resource assignment, 31
Resource sharing, 31–33, 91, 100, 115, 123, 128–131, 133–136, 140, 145, 147, 153
Retiming, 24, 116, 139
Rising delay block, 152

Scheduling, 9, 24, 30–32, 34–35, 81, 95–97, 100, 110, 115, 128–129, 135–137, 139–140, 145, 153
Selective delay insertion, 145, 151
Short-circuit power, 18, 20
Sign bits, 47, 49–50
Soft macroblocks, 43, 45
Sources of power consumption, 17–18
SPA, 46
Spatial correlation, 53–54, 57
Standard deviation, 47, 53–54, 59
State encoding, 24
Static power, 18, 21
Statistical sampling, 63–64
Strength reduction, 116
Structural RTL, 6, 9, 25, 32–33, 35, 115
Supply voltage reduction, 22–23, 117, 139

Supply voltage scaling, 23, 117, 128, 138–139
Technology mapping, 24, 29
Temporal correlation, 46–47, 53–54, 77, 131–132
Thermal considerations, 3
Training sequences, 43, 45
Transparent latch, 97–102, 107–108
Two's complement, 46

Uniform white noise, 46

Variable elimination, 54
Verilog, 26
VHDL, 8, 26, 31
Voltage level converters, 128

Word-length reduction, 116, 119